Functions of Several Variables

for use with

Written by MIROSLAV LOVRIĆ

NELSON

NELSON

Functions of Several Variables
by Miroslav Lovrić

for use with *Calculus for the Life Sciences*, Second Canadian Edition
by Frederick Adler and Miroslav Lovrić

Vice President, Editorial Higher Education:
Anne Williams

Publisher:
Paul Fam

Executive Editor:
Jackie Wood

Marketing Manager:
Leanne Newell

Developmental Editor:
Suzanne Simpson Millar

Technical Checker:
Caroline Purdy

Content Production Manager:
Claire Horsnell

Copy Editor:
Heather Sangster at Strong Finish

Design Director:
Ken Phipps

Managing Designer:
Franca Amore

Cover Design:
Martyn Schmoll

Cover Image:
iLexx/iStockphoto.come

Printed and bound in Canada
5 20 19

For more information contact Nelson Education Ltd., 1120 Birchmount Road, Toronto, Ontario, M1K 5G4. Or you can visit our Internet site at nelson.com

ISBN-13: 978-0-17-657136-8
ISBN-10: 0-17-657136-1

Functions of Several Variables

The aim of this module is to provide a comprehensive and thorough introduction to the ideas, major results, and applications of functions of several variables. Keeping in mind that this is a module for life sciences students, we do not go deep into mathematical theory (for instance, we do not discuss the precise definition of the limit or present the proof of the existence of absolute extreme values on closed and bounded sets). Instead, we focus on general results, mathematical understanding, and applications.

We approach the content from many directions (algebraic, numerical, geometric, and verbal) to facilitate the learning process, to equip the reader with a good understanding, and to present the vocabulary needed to communicate mathematical ideas and results.

What's in the module? In Section 1, we define basic ideas related to functions of several variables and introduce a few applications. We revisit these applications (and introduce many new ones) and deepen our understanding in light of the new mathematics that we learn. For instance, the first thing we introduce is the function $c(x, t)$, which models the concentration of a pollutant spreading from a single source (as an example of a diffusion process). Next, we sketch the graph of the function and learn more about diffusion by investigating properties of curves related to the graph (Section 2). We interpret the partial derivatives of $c(x, t)$ in Section 4 and study the diffusion partial differential equation in Section 8.

We discuss visual representation of functions in Section 2. Besides drawing graphs of surfaces, we learn how to use two-dimensional objects that we are familiar with (such as curves) to obtain three-dimensional information about the functions involved. Next, we talk about limits and continuity (Section 3) and construct partial derivatives in order to measure the rate of change of a function of several variables (Section 4). Using partial derivatives, we build the linear approximation (tangent plane) in Section 5. Section 6 contains technical material on the calculation of derivatives, namely the chain rule and implicit differentiation. The directional derivative (Section 9) enables us to determine how a function changes in any given direction.

Second-order partial derivatives are used to build a degree-2 Taylor polynomial (Section 7) and to introduce partial differential equations (Section 8). Partial differential equations are an essential tool in modelling life sciences (and other) phenomena mathematically.

The module closes with two sections on optimization: local and absolute extreme values (Section 10) and extreme values with constraints (Section 11).

Among the many applications discussed in the module are body surface area, wind chill index, estimation of the size of a tumour from a mammogram, dynamics of prey consumption (type-2 functional response), barometric formula for air pressure in the atmosphere, travelling waves and the wave equation, the relation between optimization and symmetry, and measurements of the diversity of species.

The approach used in writing this module — clear explanations and easy-to-understand narratives; numerous graphs, pictures and diagrams; a large number of fully solved examples and end-of-section exercises; and a wide spectrum of life sciences applications — makes the material suitable for students whose interests lie in life sciences and who are willing to deepen their understanding of life sciences phenomena.

I thank you for chosing this module, and I hope that you will like reading it and that you will benefit from it — that you will learn some good and useful math.

Miroslav Lovrić
McMaster University, 2014

[Solutions to odd-numbered exercises from this module are posted (free download) on the web page www.nelson.com/site/calculusforlifesciences.]

Outline

1 Introduction

Released from a source, a pollutant spreads into the surrounding air. The formula

$$\frac{1}{\sqrt{4\pi Dt}}\, e^{-x^2/4Dt}$$

can be used to describe (under certain conditions, such as no wind) the concentration of the pollutant at a location x units away from the source at time t (D denotes the positive *diffusion coefficient*). This formula is an example of a real-valued function of two variables, which we denote by

$$c(x,t) = \frac{1}{\sqrt{4\pi Dt}}\, e^{-x^2/4Dt}$$

The input consists of two real numbers (the location x and the time t), and the output is a real number.

The dependence of the body mass index (BMI) on a person's mass m (in kilograms) and height h (in metres) can be expressed as

$$\text{BMI}(m,h) = \frac{m}{h^2}$$

This is another example of a real-valued function of two variables.

The resistance R to the flow of blood through a vessel of length L and radius r is given by the formula (known as *Poiseuille's Law*)

$$R(L,r) = k\frac{L}{r^4}$$

The value of the constant k is determined by the viscosity of the blood.

To model the temperature of the water in a lake, we use a function $T(x,y,z,t)$ of four variables: the variables x and y specify the location (for instance, longitude and latitude), z specifies the depth at which we measure the temperature, and t is the time.

In order to study functions such as the concentration of a pollutant, the body mass index, or the resistance to the flow of blood, we need to extend the concepts of calculus (in particular, limits, continuity, and derivatives) to functions of several variables. Again, the major objective is to develop tools that will allow us to **study and interpret change.**

Basic Definitions and Notation

When working with functions of one variable, we use $y = f(x)$ to say that the variable y depends on *one* independent variable x. The domain of a function $y = f(x)$ is a subset of the real numbers.

By $\mathbb{R}^2$ we denote the set of all ordered pairs (x,y) of real numbers. Using set notation, we write

$$\mathbb{R}^2 = \{(x,y) \mid x \in \mathbb{R} \text{ and } y \in \mathbb{R}\}$$

Often, we visualize $\mathbb{R}^2$ as the set of points in the plane.

Definition 1 Real-Valued Function of Two Variables

A *real-valued function f of two variables* is a rule that assigns, to each ordered pair (x,y) in some subset D of $\mathbb{R}^2$, a unique real number z. We write

$$z = f(x,y)$$

The set D is called the *domain* of f. The set of all values z in $\mathbb{R}$ such that $z = f(x,y)$ for some ordered pair (x,y) in D is called the *range* of f.

The functions mentioned in the introduction to this section—the concentration of a pollutant, the body mass index, and the blood resistance—are functions of two variables. In order to calculate the value of a function of two variables, we need to know the values of both variables. For instance, the body mass index of a person of mass 71 kg and height 1.69 m is

$$\text{BMI}(71, 1.69) = \frac{71}{1.69^2} = 24.9$$

The concentration of a pollutant in the case when the diffusion coefficient is $D = 1$ is given by

$$c(x,t) = \frac{1}{\sqrt{4\pi t}} e^{-x^2/4t}$$

In particular, the concentration of a pollutant at a location $x = 5$ units away from the source at time $t = 2.5$ is

$$c(5, 2.5) = \frac{1}{\sqrt{4\pi(2.5)}} e^{-(5)^2/4(2.5)} = \frac{1}{\sqrt{10\pi}} e^{-2.5} \approx 0.01464.$$

To extend Definition 1 to an arbitrary number of variables, we introduce a new piece of notation: by $\mathbb{R}^n$, for $n \geq 2$, we denote the set of all ordered n-tuples $(x_1, x_2, \ldots, x_n)$ of real numbers. In particular, $\mathbb{R}^3$ represents the set of all ordered triples of real numbers

$$\mathbb{R}^3 = \{(x, y, z) \mid x \in \mathbb{R},\ y \in \mathbb{R},\ \text{and } z \in \mathbb{R}\}$$

which we visualize as three-dimensional space.

Definition 2 **Real-Valued Function of n Variables**

A *real-valued function f of n variables* is a rule that assigns a unique real number z to each ordered n-tuple $(x_1, x_2, \ldots, x_n)$ in a subset D of $\mathbb{R}^n$. We write

$$z = f(x_1, x_2, \ldots, x_n)$$

The set D is the *domain of f*, and the set of all values of z constitutes the *range of f*.

Thus, a number z is in the range of a function f if $z = f(x_1, x_2, \ldots, x_n)$ for some $(x_1, x_2, \ldots, x_n)$ in the domain of f (see Examples 1.3 and 1.4 for how to find the range of a function).

When a function depends on a small number of variables, we write $f(x, y)$, $f(x, t)$, $f(x, y, z)$, and the like, instead of naming independent variables using subscripts.

Let us clarify the meaning of the word "ordered" in "ordered pair" and "ordered n-tuple." Consider the function

$$f(x, y, t) = \frac{t^2 + x}{1 + x^2 + y^2}$$

The notation $f(x, y, t)$ lists the three variables in the *order* x, then y, then t. Thus, to calculate the value of f at $x = 2$, $y = 3$, and $t = 0.5$, we write

$$f(2, 3, 0.5) = \frac{(0.5)^2 + 2}{1 + (2)^2 + (3)^2} = \frac{2.25}{14}$$

Likewise,

$$f(0, -4, 1) = \frac{(1)^2 + 0}{1 + (0)^2 + (-4)^2} = \frac{1}{17}$$

gives the value of f when $x = 0$, $y = -4$, and $t = 1$.

Of course, the value of f depends on the order of its independent variables: if $x = -4$, $y = 0$, and $t = 1$, then

$$f(-4, 0, 1) = \frac{(1)^2 + (-4)}{1 + (-4)^2 + (0)^2} = -\frac{3}{17}$$

Example 1.1 Functions of Several Variables

The function $f(x, y) = 3x^2y - xy^4 - y - 1$ is a polynomial in two variables. Its domain consists of all ordered pairs in $\mathbb{R}^2$. A function of the form $f(x, y) = ax + by + c$, where a, b, and c are real numbers, is called a *linear function* of two variables. Its domain is $\mathbb{R}^2$ (a linear function is a special case of a polynomial in two variables).

In general, a linear function of n variables is defined by

$$f(x_1, x_2, \ldots, x_n) = a_1x_1 + a_2x_2 + \cdots + a_nx_n + b$$

where a_1, a_2, ..., a_n, and b are constants. The function

$$f(x, y, z) = \sqrt{x^2 + y^2 + z^2}$$

measures the distance from a point (x, y, z) in $\mathbb{R}^3$ to the origin. It is a real-valued function of three variables; its domain is $\mathbb{R}^3$, and its range consists of zero and all positive numbers.

Definition 3 Polynomial and Rational Functions

A *polynomial* in two variables x and y is a sum of terms of the form cx^ky^l, where c is a real number and k and l are integers such that $k, l \geq 0$. A *rational function* is a quotient of two polynomials.

For example, $f_1(x, y) = 3 - 2.5x + x^2y^3 - 6y^7$, $f_2(x, y) = x^{11} - 4xy + x^8y^7$, and $f_3(x, y) = 1 - x$ are polynomials in two variables, and

$$f_4(x, y) = \frac{x^{11} - 4xy}{x^2y^2 - y + 1} \quad \text{and} \quad f_5(x, y) = \frac{2}{x + y^4}$$

are rational functions.

Example 1.2 Domain of a Function of Two Variables

Find the domain of the rational function $f(x, y) = \dfrac{1}{xy - 1}$.

Note that if $xy - 1 = 0$, then $xy = 1$ and $y = 1/x$. Thus the points on the hyperbola $y = 1/x$ do not belong to the domain of f. In other words, the points in the plane $\mathbb{R}^2$ that do not lie on the hyperbola $y = 1/x$ form the domain of f.

In Figure 1.1 we shaded the domain of f; to say that the hyperbola $y = 1/x$ does not belong to the domain, we drew it using a dashed curve.

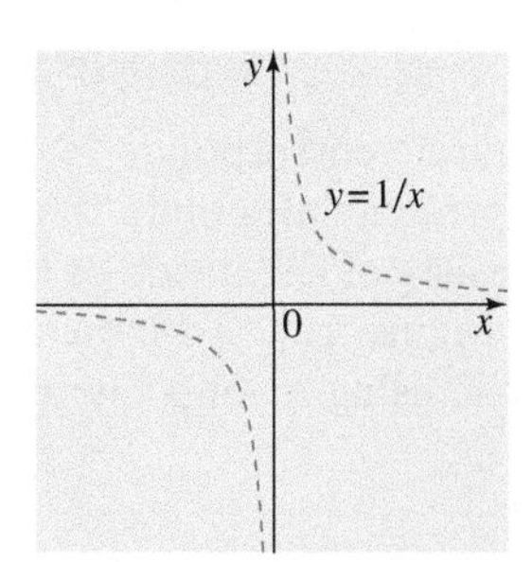

FIGURE 1.1

The domain of $f(x, y)$ from Example 1.2

Example 1.3 Range of a Function of Two Variables

Find the range of the function $f(x, y) = \dfrac{1}{xy - 1}$.

▶ The numerator of the fraction is $1 \neq 0$, and thus the range of f does not contain 0. We claim that the range (denote it by R) contains all other numbers, i.e.,

$$R = \{z \in \mathbb{R} \mid z \neq 0\}$$

To prove our claim, we need to show that no matter which $z \neq 0$ we pick, we can find a value for x and a value for y so that $f(x, y) = z$. For instance, for $z = 1$ we need to find x and y so that

$$f(x, y) = \frac{1}{xy - 1} = 1$$

Simplifying, we get $xy - 1 = 1$ and $xy = 2$. Thus, if we take $x = 1$ and $y = 2$, we get

$$f(1, 2) = \frac{1}{(1)(2) - 1} = \frac{1}{1} = 1$$

Of course, any choice for x and y, as long as $xy = 2$, will do, but keep in mind that all we need is to find *one* ordered pair (x, y) such that $f(x, y) = 1$.

Now we repeat the calculation for a general value $z \neq 0$:

$$f(x, y) = z$$

$$\frac{1}{xy - 1} = z$$

$$xy - 1 = \frac{1}{z}$$

(this is where we need $z \neq 0$)

$$xy = 1 + \frac{1}{z}$$

When $x = 1$, then $y = 1 + 1/z$ (note that y is defined as long as $z \neq 0$). To check:

$$f(1, 1 + 1/z) = \frac{1}{(1)(1 + 1/z) - 1} = \frac{1}{1/z} = z$$

▲

Example 1.4 Domain and Range of a Function of Two Variables

Find the domain and the range of the function $f(x, y) = \sqrt{16 - x^2 - y^2}$.

▶ An ordered pair (x, y) belongs to the domain of f if $16 - x^2 - y^2 \geq 0$, i.e., if $x^2 + y^2 \leq 16$.

To describe the domain geometrically, we recall that the equation $x^2 + y^2 = 16$ represents the circle of radius 4 centred at the origin. This circle divides the plane $\mathbb{R}^2$ into two regions: the inside of the circle and the outside of the circle. The points (x, y) in one of the two regions must satisfy $x^2 + y^2 < 16$; in the other, $x^2 + y^2 > 16$. To determine which is which, we use test points. The point $(0, 0)$ belongs to the inside of the circle; its coordinates satisfy

$$x^2 + y^2 = 0^2 + 0^2 = 0 < 16$$

Thus, the inside of the circle corresponds to the inequality $x^2 + y^2 < 16$. Consequently, $x^2 + y^2 > 16$ represents all points that lie outside the circle. We conclude that the domain of f consists of the inside of the circle $x^2 + y^2 = 16$, together with its boundary circle; see Figure 1.2. To indicate that the boundary circle belongs to the domain, we drew it using a solid line (compare with the use of the dashed line in Example 1.2).

Now the range: note that

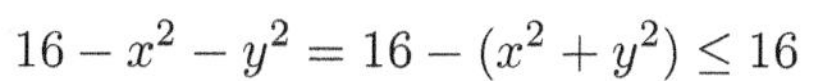

$$16 - x^2 - y^2 = 16 - (x^2 + y^2) \leq 16$$

(we are subtracting a positive number or zero from 16); therefore

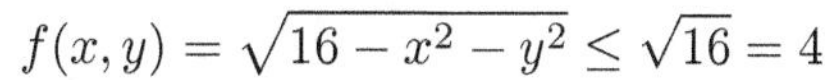

$$f(x, y) = \sqrt{16 - x^2 - y^2} \leq \sqrt{16} = 4$$

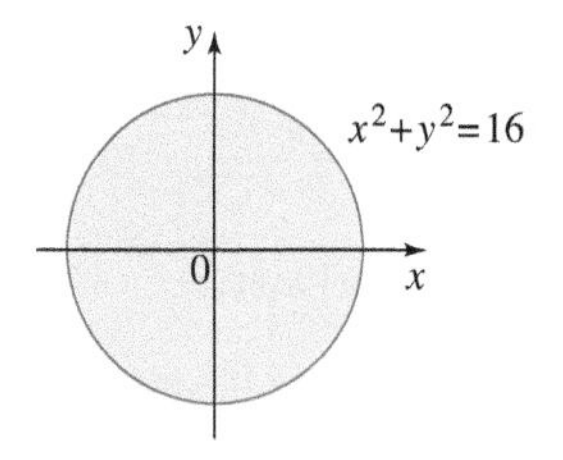

FIGURE 1.2

The domain of the function $f(x, y) = \sqrt{16 - x^2 - y^2}$

Since the square root is zero or positive, we conclude that $0 \leq f(x, y) \leq 4$, i.e., the range of f is *contained* in $[0, 4]$. To show that the range is actually *equal to* $[0, 4]$, we need to show that any number z in $[0, 4]$ can be obtained as the value of f. In other words, we need to find x and y so that $f(x, y) = \sqrt{16 - x^2 - y^2} = z$, i.e.,

$$16 - x^2 - y^2 = z^2$$

Take $x = 0$; then $16 - y^2 = z^2$, $y^2 = 16 - z^2$, and $y = \pm\sqrt{16 - z^2}$ (note that the expression under the square root is greater than or equal to zero, since z is in $[0, 4]$; thus y is a real number). We need only one y value, so we take $y = \sqrt{16 - z^2}$. It follows that

$$f\left(0, \sqrt{16 - z^2}\right) = \sqrt{16 - (0)^2 - \left(\sqrt{16 - z^2}\right)^2} = \sqrt{16 - (16 - z^2)} = \sqrt{z^2} = z$$

($\sqrt{z^2} = z$ since $z \geq 0$). ▲

The domain of a function as given in Definitions 1 and 2 is called the *natural domain* of a function. In some cases, we prescribe the domain, and then we refer to it as the *given domain*. For instance, the domain (natural domain) of $f(x, y) = e^{2x-3} + \sin y^2$ is $\mathbb{R}^2$. If we say

$$f(x, y) = e^{2x-3} + \sin y^2, \text{ where } x \geq -1 \text{ and } y > 0$$

then that's what we work with; i.e., the (given) domain of f is the set

$$D = \{(x, y) \mid x \geq -1, y > 0\}$$

In applications, it is the context that determines the domain. For instance, the domain of the body mass index function $\text{BMI}(m, h) = m/h^2$ consists of values of m that range from the lightest to the heaviest human, whereas h ranges from the shortest to the tallest human (as a matter of fact, the domain is even more restricted; it is known that the BMI does not give meaningful values for small children and tall people). Note that the *natural* domain of $\text{BMI}(m, h)$, which is

$$D = \{(m, h) \mid m, h \in \mathbb{R}, h \neq 0\}$$

makes no sense in the context of measurements of mass and height.

Example 1.5 Wind Chill

On a winter day, we might feel cold. This is because our body senses the temperature of our skin, which is exposed to the surrounding cold air. When it's windy we lose heat faster and feel that it is colder than it actually is (judging solely by the air temperature). The *wind chill* (also called the *wind chill index*) quantifies this sensation of coldness.

In 2001, Environment Canada started calculating the wind chill based on the function

$$W(T, v) = 13.12 + 0.6215T - 11.37v^{0.16} + 0.3965Tv^{0.16}$$

where T denotes the air temperature (in degrees Celsius) and v is the wind speed (in kilometres per hour), measured at 10 m above the ground. This formula gives meaningful values for temperatures below 10°C and wind speeds greater than 5 km/h. For instance,

$$W(-10, 30) = 13.12 + 0.6215(-10) - 11.37(30)^{0.16} + 0.3965(-10)(30)^{0.16} \approx -19.5.$$

Thus, when the air temperature is -10°C and the wind blows at 30 km/h, it feels as if the temperature is actually -19.5°C.

Let's try to understand why this wind chill formula makes sense. Pick a value for the temperature, say, $T = -5^{\circ}$C. Then

$$\begin{aligned} W(-5, v) &= 13.12 + 0.6215(-5) - 11.37v^{0.16} + 0.3965(-5)v^{0.16} \\ &= 10.0125 - 13.3525v^{0.16} \end{aligned}$$

So $W(-5, v)$ is a function of one variable, the wind speed v. Its derivative

$$W'(-5, v) = -13.3525(0.16)v^{-0.84} = \frac{-2.1364}{v^{0.84}}$$

is negative (keep in mind that $v > 0$); thus v is a decreasing function of v. This makes sense: the wind chill decreases (i.e., becomes more negative) as the wind blows faster (and the air temperature remains unchanged at -5°C). The function

$$\begin{aligned} W(T, 40) &= 13.12 + 0.6215T - 11.37(40)^{0.16} + 0.3965T(40)^{0.16} \\ &= -7.3959 + 1.3369T \end{aligned}$$

shows how the wind chill depends on the temperature with a constant wind speed of 40 km/h. We see that $W(T, 40)$ has positive slope; thus, as the temperature increases, so does the wind chill.

Example 1.6 Body Surface Area

The *body surface area* (BSA), i.e., the total surface area of the human body, is an important measurement used in diagnosis and treatment decisions for certain medical conditions. A good estimate of the body surface area of a cancer patient is of crucial importance in obtaining accurate dosage calculations for a chemotherapy treatment [see T. Vu. (1999). "Standardization of body surface area calculations." T. Vu, Dept. of Pharmacy, Cross Cancer Institute. Available at www.halls.md/bsa/bsaVuReport.htm].

Using statistical means, researchers have determined that the formula

$$S_D(m, h) = 0.20247m^{0.425}h^{0.725}$$

gives a good approximation of the body surface area of a human of height h (metres) and mass m (kilograms). In the literature, this formula is called the *DuBois and DuBois formula.* It says that, for instance, the BSA of a person of height 1.75 m and mass 76 kg is equal to

$$S_D(76, 1.75) = 0.20247(76)^{0.425}(1.75)^{0.725} \approx 1.914 \text{ m}^2$$

There are other estimates. For instance, the British Columbia Cancer Agency, the British Columbia Children's Hospital, and several other institutions use the *Mosteller formula*

$$S_M(m, h) = \frac{1}{6}\sqrt{mh}$$

(with height h in metres and mass m in kilograms). For a person with $m = 76$ kg and $h = 1.75$ m, we get

$$S_M(76, 1.75) = \frac{1}{6}\sqrt{(76)(1.75)} \approx 1.922 \text{ m}^2$$

which is quite close to the DuBois and DuBois estimate of 1.914.

Example 1.7 Determining the Size of a Tumour from a Mammogram

After a tumour has been detected in a mammogram, the first question that arises is, how big is it? A mammogram provides a two-dimensional (flat) image, so how do we determine its three-dimensional size, i.e., its volume?

Some tumours look (approximately) like ellipsoids. The volume of an ellipsoid with diameters a, b, and c is (Figure 1.3a)

$$V = \frac{4}{3} \cdot \frac{a}{2} \cdot \frac{b}{2} \cdot \frac{c}{2} \pi$$

The greatest distance across the flat image of a tumour is taken as one diameter (say, a); see Figure 1.3b. The diameter b is the greatest distance perpendicular to a. For the third diameter (the one that does not show in the flat image), it is common practice to take the average of a and b, so $c = (a + b)/2$.

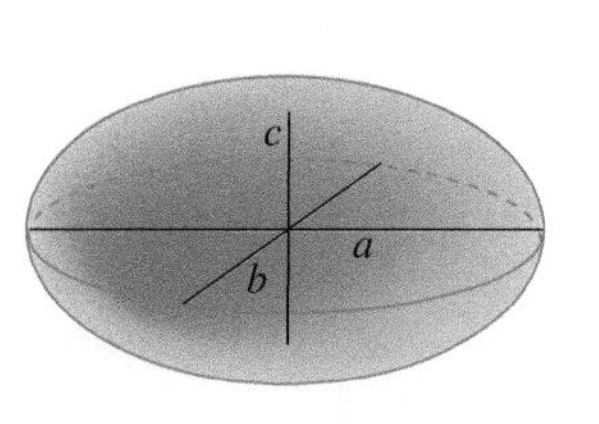

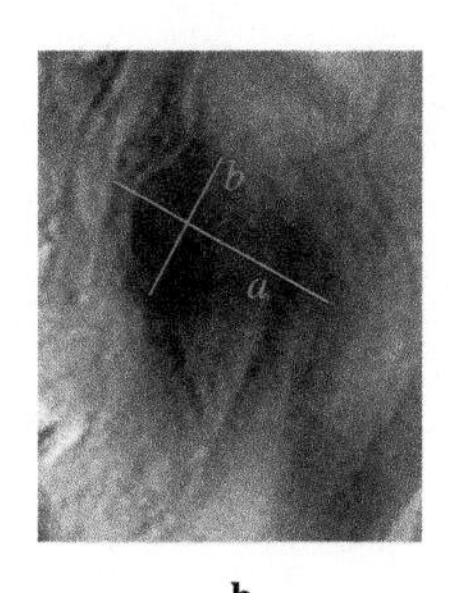

a b

FIGURE 1.3

Estimating the size of a tumour from a mammogram

(Photo courtesy of Miroslav Lovrić)

Thus, the function of two variables

$$V(a, b) = \frac{4}{3} \cdot \frac{a}{2} \cdot \frac{b}{2} \cdot \frac{(a + b)/2}{2} \pi = \frac{ab(a + b)}{12} \pi$$

gives a simple and useful estimate of the volume of the tumour. Based on information about density (number of cells per unit volume), one can obtain an approximation for the number of tumour cells.

Is this how it's really done?

Modern imaging techniques produce a sequence of images — parallel cross-sections of the tumour taken, say, d units apart (Figure 1.4a). Computer software is used to superimpose a square grid over each image and count those squares ("pixels") that are within or on the boundary of the tumour; see Figure 1.4c.

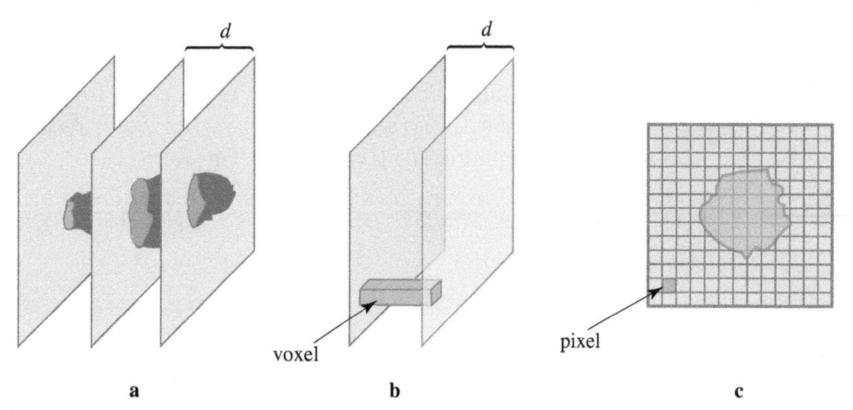

FIGURE 1.4

Imaging cancer

Each of these pixels is then thought of as a three-dimensional box (called a *voxel*; Figure 1.4b) whose volume is known (because the grid size and d are known). Multiplying the number of voxels by the volume of one voxel gives an estimate for the volume of the tumour.

If you are not familiar with the three-dimensional coordinate system, read the Appendix at the end of this section before you proceed.

Example 1.8 Domain of a Function of Three Variables

Find, and describe geometrically, the domain of the function $f(x, y, z) = \ln z$.

A point (x, y, z) belongs to the domain D of f if $z > 0$. Thus,

$$D = \{(x, y, z) \mid x \in \mathbb{R}, y \in \mathbb{R}, z > 0\}$$

The set D consists of all points in $\mathbb{R}^3$ whose z-coordinate is positive. It is the upper half-space, i.e., the set of all points in space that lie above the xy-plane.

Example 1.9 Domain of a Function of Three Variables

Find the domain of the function $f(x, y, z) = \sqrt{xyz}$.

A point (x, y, z) is in the domain of f if $xyz \geq 0$. The equations $x = 0$, $y = 0$, and $z = 0$ describe the three coordinate planes. If $xyz > 0$, then either all three coordinates are positive, or exactly two coordinates are negative. The locations of these points are given in Table 1.1.

Table 1.1

x	y	z	Octant
+	+	+	1st
−	−	+	3rd
−	+	−	6th
+	−	−	8th

Thus, D consists of the first, third, sixth, and eighth octants, including the three coordinate planes.

We will spend most of our time studying functions of *two* variables. Although conceptually there are no significant differences between functions of 2 and 25 variables, technical and computational difficulties increase significantly with the number of variables. For example, a function of two variables has three second derivatives (actually four, in the worst case); a function of 25 variables has 325 second derivatives (625 in the worst case).

In this book we focus on *real-valued* functions. A *vector-valued function* is a function whose range is a subset of $\mathbb{R}^n$, with $n \geq 2$. For example, the range of the function

$$f(x, y) = (x^3 y, e^y + x^2)$$

is contained in $\mathbb{R}^2$, and so its values can be visualized as vectors in a plane. In particular,

$$f(2, 0) = (2^3(0), e^0 + 2^2) = (0, 5)$$

and

$$f(-3, 1) = ((-3)^3(1), e^1 + (-3)^2) = (-27, e + 9)$$

We will briefly mention a few facts about vector functions at the end of Section 2. In linear algebra we study *linear transformations*, which are special cases of vector-valued functions. Vector-valued functions are usually studied in advanced calculus courses (which are usually referred to as "Vector Calculus").

Appendix: Three-Dimensional Cartesian Coordinate System

To build a three-dimensional (Cartesian) coordinate system, we use three mutually perpendicular number lines, called the *coordinate axes*, placed so that they intersect at a point (called the origin) that represents the number zero for all three axes. The coordinate axes are identified as the x-axis, the y-axis, and the z-axis. The pairs of coordinate lines form three *coordinate planes:* the xy-plane, the yz-plane, and the xz-plane (see Figure 1.5a).

The coordinate axes divide the plane $\mathbb{R}^2$ into four quadrants. Moving one dimension higher, we say that the three coordinate planes divide the space $\mathbb{R}^3$ into eight octants, defined (as the quadrants in $\mathbb{R}^2$) based on the signs of the coordinates, as shown in Table 1.2.

Table 1.2

x	y	z	Location	Octant
+	+	+	top-front-right	1st
−	+	+	top-back-right	2nd
−	−	+	top-back-left	3rd
+	−	+	top-front-left	4th
+	+	−	bottom-front-right	5th
−	+	−	bottom-back-right	6th
−	−	−	bottom-back-left	7th
+	−	−	bottom-front-left	8th

The location of a point P in space is uniquely identified using the following three numbers: the directed distance p_1 from P to the yz-plane, called the x-coordinate of P; the directed distance p_2 from P to the xz-plane, called the y-coordinate of P; and the directed distance p_3 from P to the xy-plane, called the z-coordinate of P (see Figure 1.5b). "Directed distance" means the usual distance, to which we assign the sign + or − depending on the location of P, according to Table 1.2.

We write $P(p_1, p_2, p_3)$ and say that the point P has coordinates (p_1, p_2, p_3). The origin has coordinates $(0, 0, 0)$.

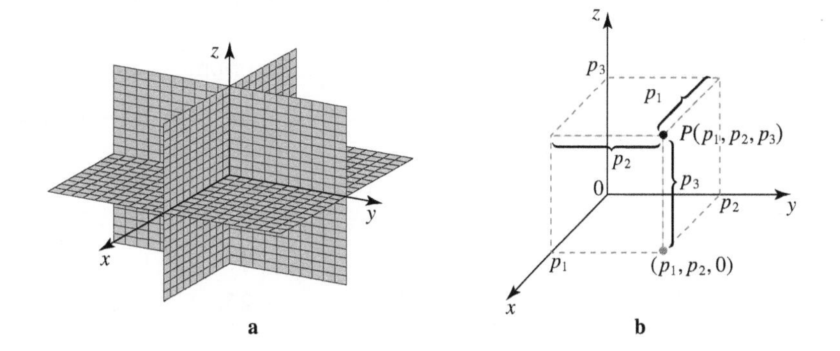

FIGURE 1.5

Three-dimensional Cartesian coordinate system

The *distance* between points $P(p_1, p_2, p_3)$ and $Q(q_1, q_2, q_3)$ in $\mathbb{R}^3$ is given by

$$d(P, Q) = \sqrt{(q_1 - p_1)^2 + (q_2 - p_2)^2 + (q_3 - p_3)^2}$$

Summary Using **functions of several variables,** we can study phenomena that depend on more than one variable. Extending the concepts from one-variable calculus, we discussed the **domain** and the **range** of a function. Examples of functions of several variables include **wind chill index, body surface area,** and **diffusion (spread) of a pollutant** as well as a formula that allows us to estimate the **size of a tumour from a mammogram.** Functions of two and three variables require that we work in the **three-dimensional Cartesian coordinate system.**

1 Exercises

1. Which of the functions $f(x,y) = 3 - x + 6y$, $g(x,y) = 2xy$, $h(x,y) = 3x - 2y + 4$, and $k(x,y) = 1$ are linear?

2. What is the domain of the function $f(x) = 1/x$? What is the domain of the function $f(x,y) = 1/x$?

3. Find a formula for a function $z = f(x,y)$ of two variables whose domain is the set $D = \{(x,y) \mid x > 0 \text{ and } y > 0\}$.

4. Find a formula for a function $z = f(x,y)$ of two variables whose domain is the set $D = \{(x,y) \mid x \neq 0 \text{ and } y \neq 0\}$.

5. Write a formula for a real-valued function f of two variables whose value at (x,y) is inversely proportional to the square of the distance from (x,y) to the origin, so that $f(-1,2) = 4$.

6. Write a formula for a real-valued function h that is proportional to the distance and inversely proportional to the square of time. It is equal to 45 when the distance is 10 m and the time is 3 s.

7–12 ▪ Find and sketch (or describe) the domain of each function.

7. $f(x,y) = xy + 1/x$

8. $g(x,y) = 5 - x - \sqrt[3]{x-y}$

9. $f(x,y) = \sqrt{5-x-y}$

10. $f(x,y) = (x^2+y^2)^{-3}$

11. $f(x,y) = (x^2+y^2-1)^{-2}$

12. $f(x,y) = \arctan(x/y)$

13. What is the domain of the concentration of the pollutant function $c(x,t)$ mentioned in the introduction to this section?

14. What is the range of the linear function $f(x,y) = 2x - 4y - 10$? Prove your claim.

15. Discuss how the range of the linear function $f(x,y) = ax + by + c$ depends on the values of a, b, and c.

16–25 ▪ Find the domain and the range of each function.

16. $f(x,y) = 5$

17. $f(x,y) = x + 12$

18. $f(x,y) = 1/(x+y)$

19. $g(x,y) = e^{x^2+y^2}$

20. $h(x,y) = \sqrt{x+y}$

21. $g(x,y) = \sqrt{2+x^2+5y^2}$

22. $f(x,y) = (x^2-y^2)^{-1}$

23. $g(x,y) = 3|x| + |y|$

24. $f(x,y) = \ln(xy)$

25. $f(x,y) = \arctan(x^2)$

26. Describe the domain of the function $f(x,y,z) = 1/x - 1/y - 1/z$.

27. What is the range of the function $f(x,y,z) = e^{-2x^2-3y^2-z^2}$?

28. Find the domain and the range of the function $g(x,y,z) = \ln(16 - x^2 - y^2 - z^2)$.

29. Find a formula for a function f whose value at (x, y, z) is proportional to the distance between (x, y, z) and $(3, 2, -4)$, and is such that $f(1, 1, 3) = 18$.

30. Sketch (or describe) the domain of $f(x, y) = \tan x \tan y$.

31. Sketch (or describe) the domain of $f(x, y) = \sqrt{\tan x \tan y}$.

32. An alternative model for the wind chill index is given by

$$W_1(T, v) = 91 + (0.44 + 0.325\sqrt{v} - 0.023v)(T - 91)$$

where T denotes the temperature (in degrees Fahrenheit) without wind and v is the wind speed in miles per hour, $5 \leq v \leq 45$.

(a) Compute $W_1(0, v)$, $5 \leq v \leq 45$, and show that it is a decreasing function. Explain why this fact makes sense.

(b) Compute $W_1(T, 10)$ and show that it is an increasing function. Explain why this fact makes sense.

33. Recall that a person of mass m (in kilograms) and height h (in metres) has a body mass index of $\text{BMI} = m/h^2$. What is the body mass index of a person who is 10% heavier and 10% taller?

34. If we wish to calculate the body mass index of a person whose weight is in pounds and height is in inches, we can no longer use $\text{BMI} = m/h^2$. Find the formula for the BMI in that situation.

35. Consider the formula $S_D(m, h) = 0.20247m^{0.425}h^{0.725}$ for the body surface area of a human of height h (metres) and mass m (kilograms), introduced in Example 1.6. What makes S_D larger: a 10% increase in mass and a 5% increase in height, or a 5% increase in mass and a 10% increase in height?

2 Graph of a Function of Several Variables

The ability to **visualize** functions helps us understand their behaviour better and more deeply. For instance, the formula $f(x, y) = \sqrt{x^2 + y^2}$ tells only one side of the story of this function. Thinking about it geometrically — as the distance between a point (x, y) in the plane and the origin, or as a cone in three-dimensional space — we can say a lot more about its properties and behaviour (for instance, we can easily deduce that it has a minimum and no maximum).

The **graph** of a function of two variables is a **surface in space.** We draw a number of surfaces in an attempt to learn to link their geometric features with their algebraic description. To further facilitate our understanding, we study various curves related to the surfaces — the most important example of which are the **contour (or level) curves.**

At the end of the section, we introduce **vector fields.** This is an important concept within mathematics (it will help us understand extreme values of functions better) and in applications (for instance, to describe all kinds of flows).

Graph of a Function of Two Variables

The graph of a function $y = f(x)$ of one variable is a curve in the xy-plane. Each point (x, y) on that curve contains two pieces of data: the value of the independent variable x and the corresponding value $y = f(x)$ of the function. Formally, we define the graph of f as the set of all ordered pairs (x, y) (i.e., points in the xy-plane) such that $y = f(x)$ for some x in the domain of f. Expanding this idea, we obtain the following definition.

Definition 4 Graph of a Function of Two Variables

The *graph* of a function $z = f(x, y)$ of two variables is the set of points (x, y, z) in the space $\mathbb{R}^3$ such that $z = f(x, y)$ for some (x, y) in the domain of f.

In order to visually represent a function $z = f(x, y)$, we need three axes (three dimensions): two coordinate axes (usually, the x-axis and the y-axis) are used for the domain, and the third one (the z-axis) is used for the range. The graph of a function of two variables is a surface in space; see Figure 2.1.

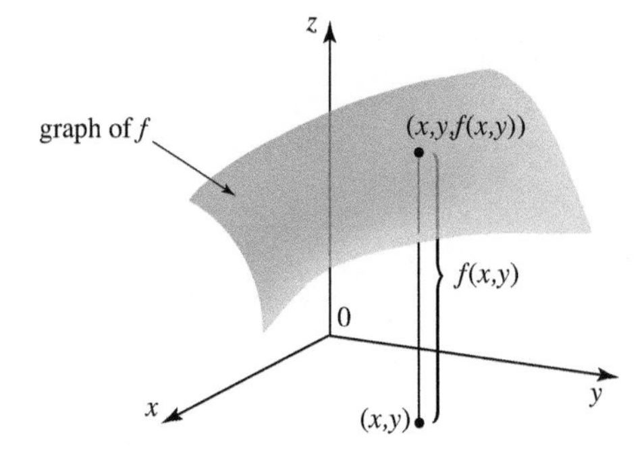

FIGURE 2.1
The graph of a function of two variables

By generalizing Definition 4 we can define the graph of a function of any number of variables. However, if a function depends on more than two variables, we have no good way of visualizing its graph; for instance, to graph a function of three variables we need four coordinate axes (i.e., four dimensions).

We now explore graphs of functions and equations involving two variables.

Example 2.1 Graph of a Constant Function

Describe the graph of the constant function $f(x, y) = C$, where C is a real number.

▶ The graph of f consists of all points whose coordinates are (x, y, C), where x and y are real numbers. Thinking of the z-coordinate as height, we see that all points (x, y, C) belong to the plane that is parallel to the xy-plane and placed C units above it (if C is positive) and $|C|$ units below it if C is negative. If $C = 0$, then the graph of $f(x, y) = 0$ coincides with the xy-plane. ▲

Arguing in a similar way, we can write equations for all planes parallel to the coordinate planes; see Table 2.1, where C represents a constant.

Table 2.1

Equation	Graph
$x = 0$	yz-plane
$x = C$	plane, parallel to the yz-plane, crossing the x-axis at $(C, 0, 0)$
$y = 0$	xz-plane
$y = C$	plane, parallel to the xz-plane, crossing the y-axis at $(0, C, 0)$
$z = 0$	xy-plane
$z = C$	plane, parallel to the xy-plane, crossing the z-axis at $(0, 0, C)$

Example 2.2 Graph of a Linear Function

Sketch the graph of the function $f(x, y) = 12 - 2x - 3y$.

▶ The function $z = f(x, y) = 12 - 2x - 3y$ is linear, and its graph is a plane. We calculate the intercepts: when $x = y = 0$, then $z = 12$; so, the point $(0, 0, 12)$ is the z-intercept. Likewise, when $x = z = 0$ then $y = 4$, and so $(0, 4, 0)$ is the y-intercept. Substituting $y = z = 0$ into f, we obtain the x-intercept $(6, 0, 0)$.

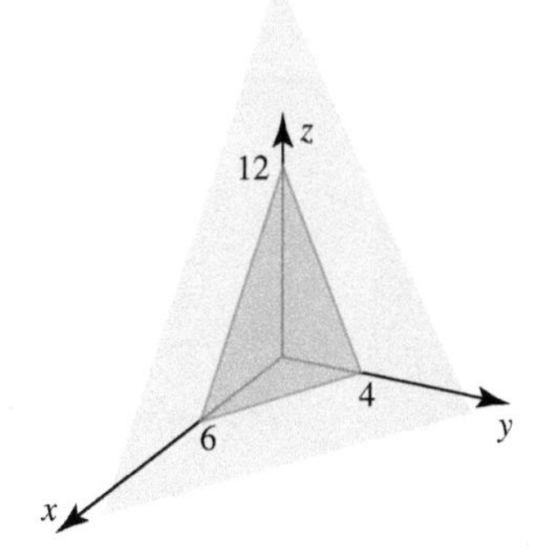

FIGURE 2.2

The graph of the linear function $f(x, y) = 12 - 2x - 3y$

The graph of f is drawn in Figure 2.2. We draw the intercepts and connect them with line segments. The triangle thus formed is the part of the plane in the first octant. ▲

Example 2.3 Graph of a Linear Function

Sketch the graph of the function $f(x, y) = -2x$.

▶ The function f is linear, so its graph must be a plane. We note that y is missing from the formula for f. How does that affect the graph?

Take $y = 0$; the graph of $z = -2x$ is a line through the origin in the xz-plane of slope -2. Since z does not depend on y, its graph looks the same no matter what y is. Thus, to graph $z = -2x$ we draw the line $z = -2x$ in all planes parallel to the xz-plane; see Figure 2.3.

We think of the graph of f as obtained by translating the line $z = -2x$ along the y-axis. ▲

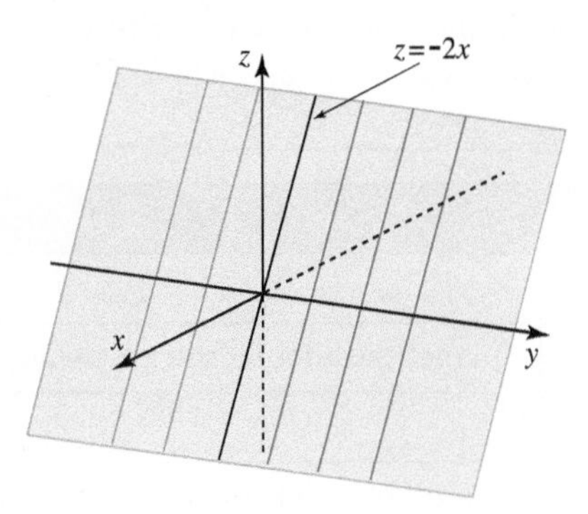

FIGURE 2.3
The graph of $f(x, y) = -2x$

Sometimes, curves are defined implicitly (using an equation). Surfaces can be defined in the same way, as our next example shows.

Example 2.4 Graph of an Equation

Sketch the graph of the equation $x^2 + y^2 = 25$ in $\mathbb{R}^3$.

▶ In $\mathbb{R}^2$, the equation $x^2 + y^2 = 25$ represents a circle of radius 5 with centre at the origin. The fact that there is no z means that the surface looks the same no matter what the height z is (compare with the case of the missing variable in Example 2.3). In other words, the surface is built of identical circles placed on top of each other (Figure 2.4). It is a cylinder of radius 5 whose axis of symmetry (also called axis of rotation) is the z-axis. ▲

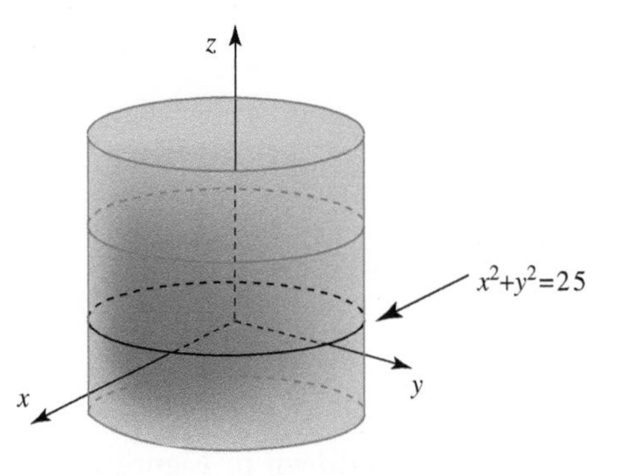

FIGURE 2.4
The cylinder $x^2 + y^2 = 25$

In general, it is difficult to draw graphs of functions of two variables by hand. We often use a graphing calculator or software (such as Maple or MATLAB) instead. In a moment, we will look into other ways of visualizing and thinking about graphs.

Example 2.5 Graphs of Functions of Two Variables

Identify the graphs of the functions $f(x,y) = x^2 + y^3$, $g(x,y) = e^{-x^2-y^2}$, and $h(x,y) = \sin x \cos y$ in Figure 2.5.

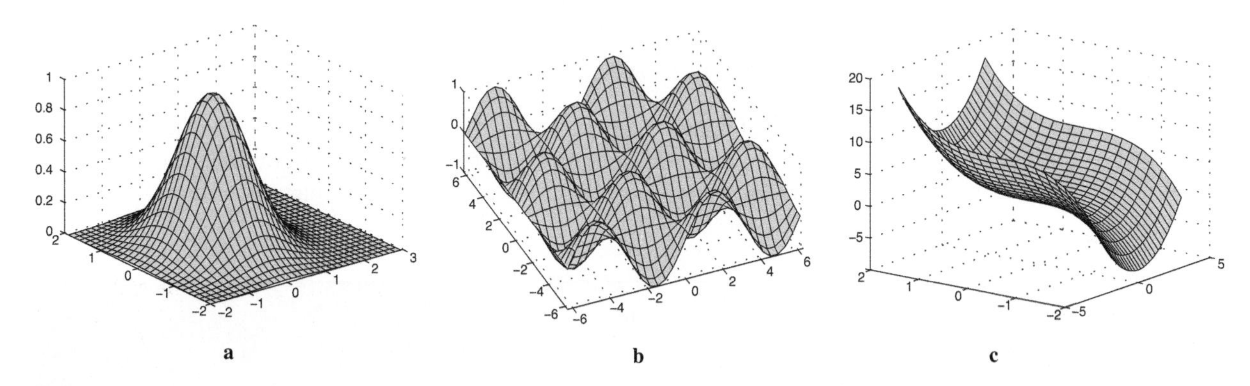

FIGURE 2.5

Graphs for Example 2.5

The "waves" in (b) repeat, suggesting that thay came from periodic functions. Thus, we claim that (b) is the graph of $h(x,y)$. To support our claim, since both $\sin x$ and $\cos y$ have values between -1 and 1, the range of h must be between -1 and 1 as well. The function in (b) is the only function with that range.

Let's think about $f(x,y) = x^2 + y^3$. When $y = 0$, then $f(x,y) = x^2$, so the graph of f increases as x approaches ∞ and $-\infty$ (and y is kept at zero). When $x = 0$, $f(x,y) = y^3$, so the graph increases as y increases and decreases as y decreases (and x is kept at zero). Thus, the graph in (c) represents f. Simply put, the function f assumes arbitrarily large positive (and arbitrarily large negative) values. Of all three graphs, (c) is only one with that property.

Note that $-x^2 - y^2 \le 0$, and thus $g(x,y) = e^{-x^2-y^2} \le e^0 = 1$; i.e., the graph does not go higher than $z = 1$ (in other words, g has a maximum at $(0,0)$, with value $g(0,0) = 1$). Since the exponential function is always positive, the graph does not touch the xy-plane or go below it. In other words, its range is $(0,1]$, so it must be (a).

Figure 2.6 shows computer-generated plots of the functions $g(x,y) = xy(x^2+y^2)^{-1}$ and $h(x,y) = \arctan(y/x)$. Bad news: even computers cannot fully reveal certain features, such as the behaviour of g at and near $(0,0)$—note that, for instance, the graph does not clearly indicate whether or not g is defined at $(0,0)$. In Section 3 we show that g does not have a limit at $(0,0)$, which cannot be seen from the graph. The graph of h is shown as a continuous surface, which is not true (we will get back to this in Section 3 as well).

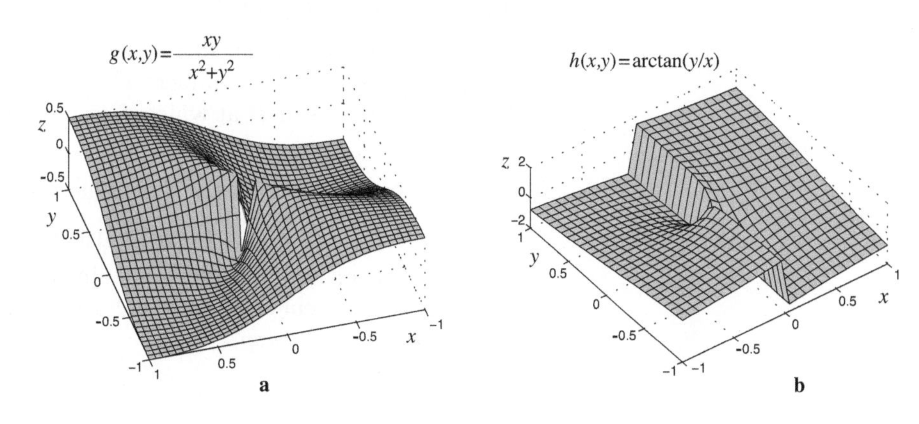

FIGURE 2.6

The graphs of the functions $g(x,y)$ and $h(x,y)$

Luckily, there is an another way of visualizing graphs of surfaces. The idea is to use *curves* on surfaces to extract information about surfaces. In this way, we reduce what is genuinely a three-dimensional situation to a two-dimensional situation (analyzing and interpreting curves).

Contour Curves (Level Curves)

A powerful way of analyzing graphs of surfaces is to draw *level curves* (or *contour curves*). This is a common idea—for instance, on a topographic map, a contour curve connects locations (points) of the same altitude (elevation above sea level); see Figure 2.7a.

a

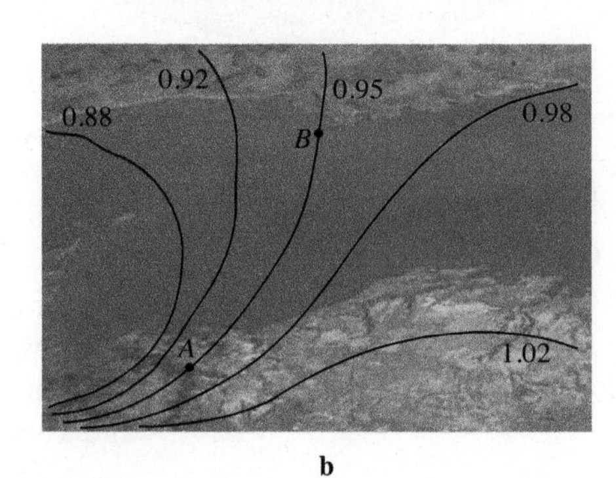

b

FIGURE 2.7

Contour curves on a topographic map, and showing air pressure

(Photo courtesy of Miroslav Lovrić)

For instance, at the point M the elevation is 120 m. As well, the elevation at all points on the contour curve that goes through M is 120 m (we call it a *contour curve of value* 120 or a *level curve of value* 120).

By looking at the topographic map in Figure 2.7a, we can not only find the elevation (or approximate elevation for locations that lie between contour curves) but deduce other important features. For instance, the hill is steeper along the trail AB than along the trail CD: the closer the contour curves are, the steeper the hill; the farther apart they are, the smaller the slopes.

We use various curves on weather maps, such as *isobars,* to show locations with the same atmospheric pressure; *isotherms,* to show locations with the same air temperature; and *isomers,* to show locations with the same monthly or seasonal precipitation.

What do the isobars in Figure 2.7b tell us? (The numerical values in the picture represent air pressure in atmospheres.) Wind—the motion of the air from regions of higher pressure towards regions of lower pressure—is stronger at A (the level curves are closer, so the pressure change happens over a smaller distance) than at B (the level curves are farther apart).

Definition 5 Contour Curve (Level Curve) of a Function of Two Variables

Let $f: D \to \mathbb{R}^2$ be a real-valued function of two variables defined on the domain $D \subseteq \mathbb{R}^2$, and let c be a real number. The *contour curve* (or *level curve*) of value c is the set of all points in D at which f assumes the value c.

In symbols, the contour curve of f of value c is the set

$$\{(x, y) \in D \mid f(x, y) = c\}$$

By definition, a contour curve is contained in the domain of the function. A collection of contour curves corresponding to different values of c forms a *contour diagram* (or a *contour map*).

Although the domain and the range of a vector field are subsets of the same set, we think of them as different objects. The elements of the domain are thought of as points, whereas the values of the vector field (i.e., the range) are visualized as vectors. To plot a vector field, we draw a vector $\mathbf{F}(x, y)$ whose initial point is (x, y); see Figure 2.16a.

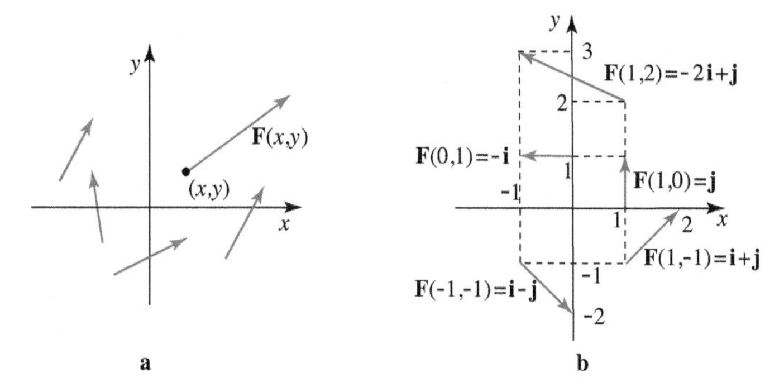

FIGURE 2.16
Graphing a vector field

Consider the vector field $\mathbf{F}(x, y) = -y\mathbf{i} + x\mathbf{j}$. To visualize the value $\mathbf{F}(1, 2) = -2\mathbf{i} + \mathbf{j}$, we draw the vector $-2\mathbf{i} + \mathbf{j}$ starting from $(1, 2)$. In the same way, we drew $\mathbf{F}(1, 0) = \mathbf{j}$, $\mathbf{F}(-1, -1) = \mathbf{i} - \mathbf{j}$, $\mathbf{F}(1, -1) = \mathbf{i} + \mathbf{j}$, and $\mathbf{F}(0, 1) = -\mathbf{i}$ in Figure 2.16b.

Using a computer to keep doing what we started by hand, we obtain the graph of $\mathbf{F}(x, y) = -y\mathbf{i} + x\mathbf{j}$ shown in Figure 2.17a. The graph of the vector field $\mathbf{F}(x, y, z) = (z, x, y)$ in $\mathbb{R}^3$ is drawn in Figure 2.17b.

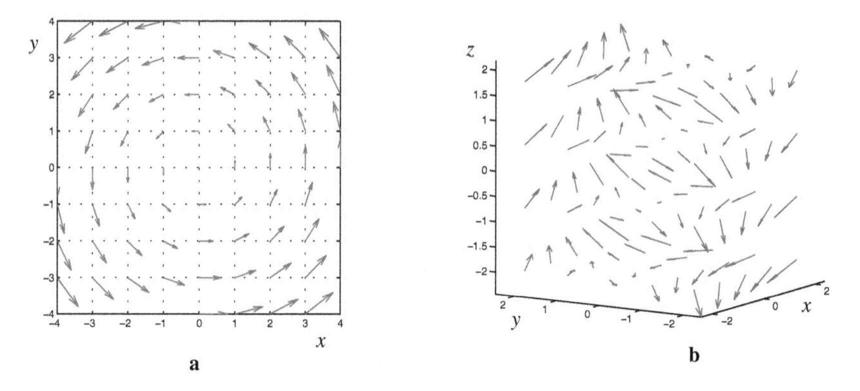

FIGURE 2.17
Graphs of vector fields

Vector fields are used, for instance, when we try to model some kind of flow (water, air, blood, etc.). In Figure 2.18a vectors are used to describe the motion of chunks of ice that have separated from a glacier. The vector field

$$\mathbf{H}(x, y) = -\frac{x + y}{x^2 + y^2}\mathbf{i} + \frac{x - y}{x^2 + y^2}\mathbf{j}$$

(Figure 2.18b) could be used to model the motion of air near the centre of a hurricane.

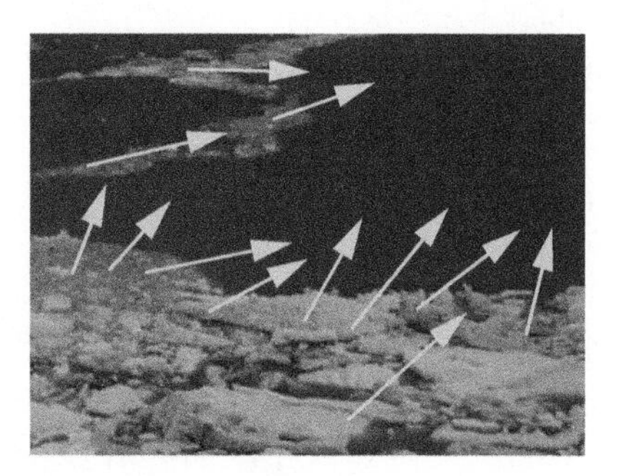

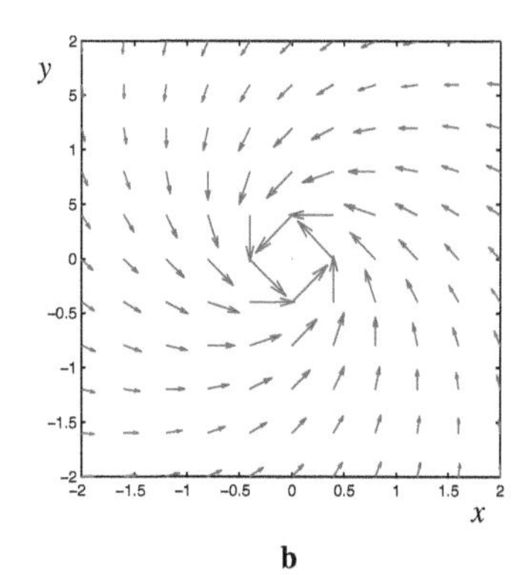

a b

FIGURE 2.18
Vector fields modelling flow
(Photo courtesy of Miroslav Lovrić)

We will meet vector fields again when we study extreme values of functions in Section 10. A special vector field (the gradient vector field) will help us understand the behaviour of a function at its critical points.

Summary The **graph** of a function or an equation involving two variables is a **surface in space.** For instance, **linear functions** are represented by **planes,** and the graph of $x^2 + y^2 = 1$ is a cylinder. We learn how to analyze surfaces by linking their **algebraic description** (formula) to their **geometric properties**. The **contour diagram** is an important tool for visually representing and interpreting functions. Used in topographic maps and all kinds of weather-related maps, a contour diagram uses two-dimensional curves to convey three-dimensional information. By analyzing various curves on surfaces (for instance intersections of the surface with vertical planes), we are able to understand the somewhat complicated behaviour of the **diffusion process.** A **vector field** is a function whose domain and range have the same dimension. All kinds of phenomena (motion of air or fluid, motion of sea ice, etc.) can be described using vector fields.

2 Exercises

1. What is a level curve? Is it possible for the level curves of two different values to intersect each other? Why or why not?

2. We have already seen that a level curve does not have to be a curve (it could be an empty set or a single point; see Example 2.7). What are the level curves of the constant function $f(x, y) = 1$?

3. Which of the contour diagrams below represents a linear function?

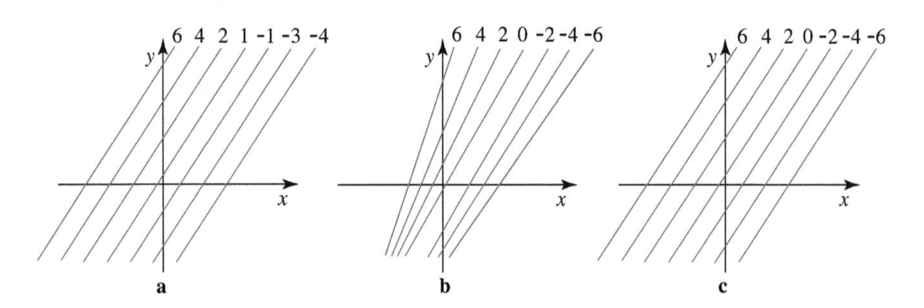

4. The contour diagrams of both the paraboloid $z = x^2 + y^2$ and the cone $z = \sqrt{x^2 + y^2}$ consist of concentric circles. What is the difference?

5. Give an example of a non-linear function whose level curves are parallel lines.

6. Describe the graph of the function $f(x, y) = x + 1$.

7. Describe the level curves of the function $g(x, y) = \sqrt{16 - x^2 - y^2}$. Sketch a contour diagram containing five to six level curves.

8–15 ▪ Describe the level curves of each function.

8. $f(x, y) = 2x + y - 7$

9. $f(x, y) = x - 4$

10. $g(x, y) = 5 - x^2 - y^2$

11. $f(x, y) = ye^x$

12. $g(x, y) = \ln(xy)$

13. $f(x, y) = y - \cos x$

14. $g(x, y) = x/y$

15. $f(x, y) = \sqrt[3]{xy}$

16–23 ▪ Draw or describe the contour diagram of each function. Use the information you obtained to sketch the graph. If necessary, compute the intersections of the surface with the yz-plane and the xz-plane.

16. $f(x, y) = x - y - 8$

17. $g(x, y) = |x|$

18. $f(x, y) = x^2$

19. $f(x, y) = 12 - x^2$

20. $z^2 = x^2 + y^2$

21. $f(x, y) = x - y - 8$

22. $f(x, y) = \sin x$

23. $f(x, y) = e^{x^2+y^2}$

24. Match each function with its contour diagram and with its graph: $f(x, y) = e^{-x^2-2y^2}$, $g(x, y) = 2xye^{-x^2-y^2}$, $h(x, y) = x/(x^2 + y^2 + 4)$, and $k(x, y) = \cos(x^2 + y^2)$.

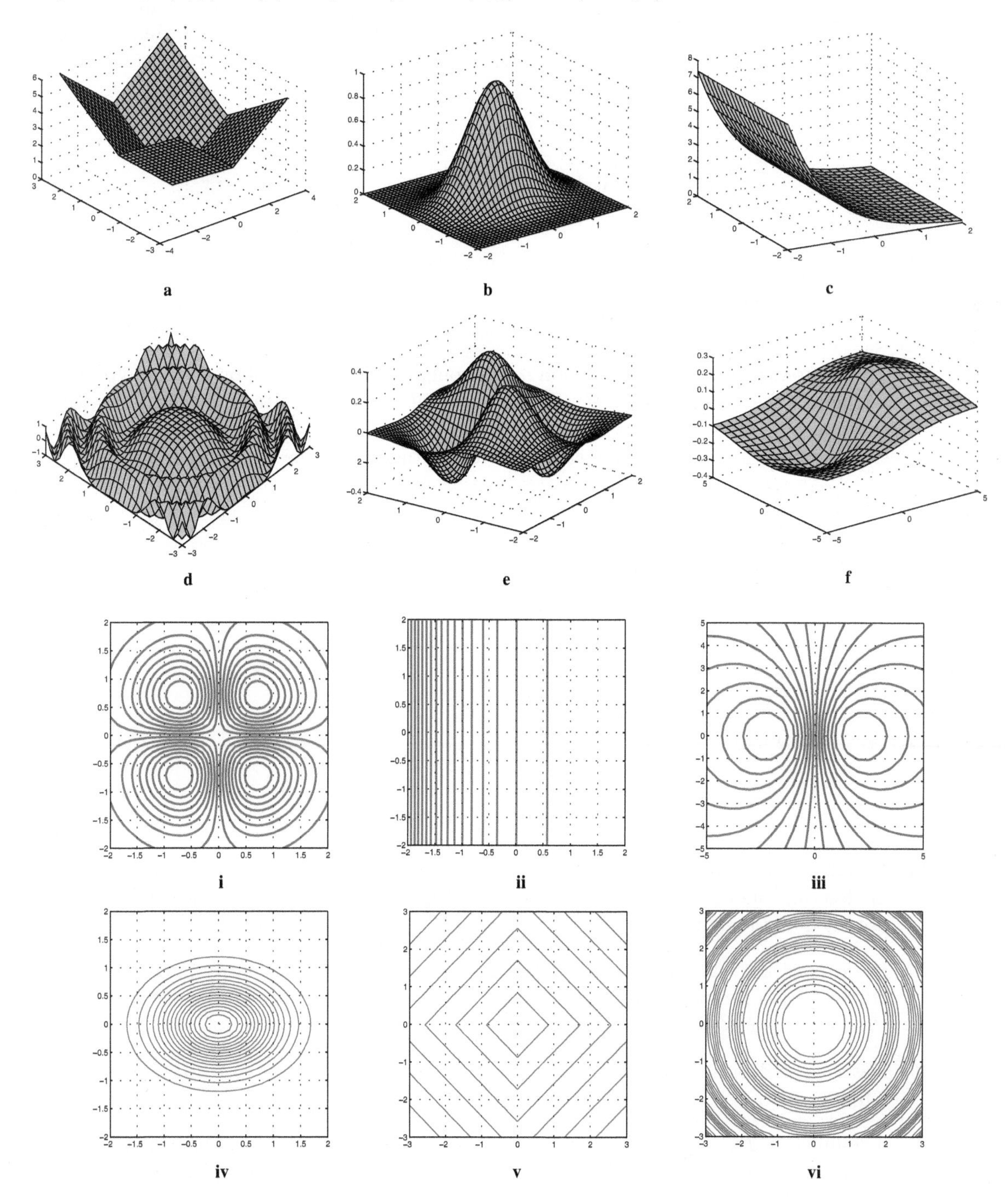

25. Draw several level surfaces of the function $f(x, y, z) = 3x + y + z$.

26. What are the level surfaces of the function $f(x, y, z) = x^2 + y^2 + z^2 - 1$?

27–30 ▪ Sketch or describe the graph of each equation in $\mathbb{R}^3$.

27. $x + 3z = 4$

28. $y = -x$

29. $x^2 + y^2 - 10 = 0$

30. $x^2 + 2y^2 + 4z^2 = 16$

31. Explain why the equation $(x - a)^2 + (y - b)^2 + (z - c)^2 = r^2$ represents a sphere of radius r centred at (a, b, c). [Hint: Start with the formula for the distance between two points in $\mathbb{R}^3$.]

32. Consider the contour diagram of the function $T(x, t)$, which models the temperature at a location x (where $0 \leq x \leq 10$) at time t (where $0 \leq t \leq 5$). Sketch the graph showing how the temperature depends on time at the location $x_0 = 7$. Sketch the graph of temperature as a function of location when $t_0 = 1$.

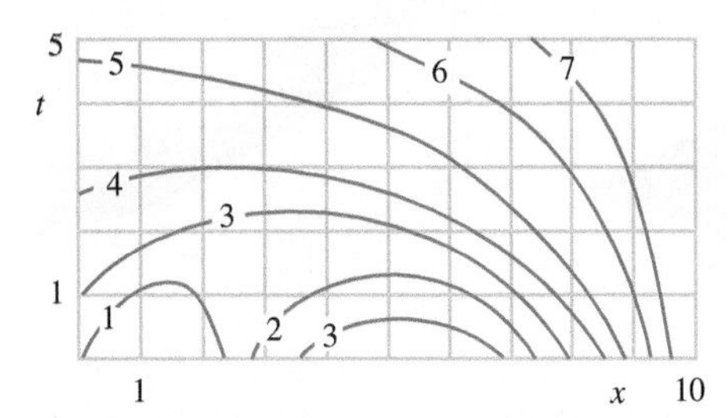

33. The contour diagram is a simulation of the concentration $c(x, y)$ of oil after a hypothetical oil spill in a region of the Atlantic Ocean south of Newfoundland (x is longitude, y is latitude). Sketch the graph showing how the concentration changes with longitude when the latitude is 44°. Sketch the graph showing how the concentration changes with latitude when the longitude is 55°.

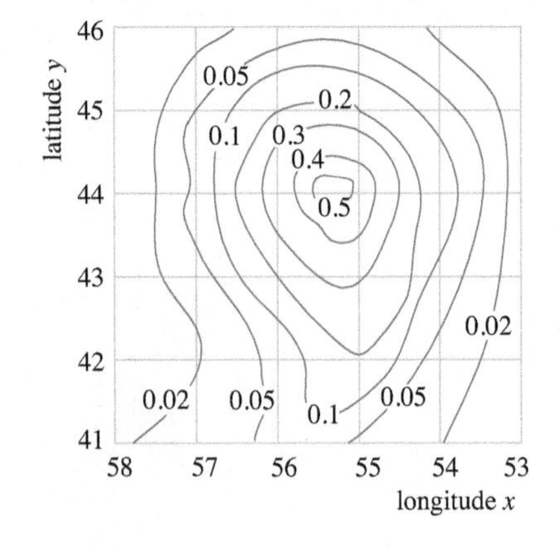

34. Switch the axes in Example 2.10; i.e., express the level curves of the BMI as functions of h. Sketch or describe the contour diagram thus obtained. What is the relation between the level curves you obtained and the curves shown in Figure 2.13b?

35. In the contour diagram of the wind chill index (see Figure 2.12), identify the five points on the contour curves that correspond to the speed of $v = 30$ km/h. They appear to be evenly spaced. Is this true? Why or why not?

36. Describe the contour diagram of the body surface area $S_D(m, h) = 0.20247m^{0.425}h^{0.725}$ of Example 1.6 in Section 1.

37. The function $T(x, t) = e^{-t}\cos x$ represents the temperature at a location x at time $t \geq 0$. Sketch the graphs of the curves $T(x, t_0)$ for $t_0 = 0$, $t_0 = 1$, and $t_0 = 5$ and explain their meaning. Sketch the graphs of the curves $T(x_0, t)$ for $x_0 = 0$, $x_0 = \pi/4$, and $x_0 = \pi$ and explain their meaning. Describe how $T(x, t)$ changes.

38. Consider the function $g(x, y) = \sin x \cos y$. Sketch the graphs of the curves $g(x, y_0)$ for $y_0 = 0$, $y_0 = \pi/2$, and $y_0 = \pi$ and explain their meaning. Sketch the graphs of the curves $g(x_0, y)$ for $x_0 = 0$, $x_0 = \pi/2$, and $x_0 = \pi$ and explain their meaning. What could $g(x, y)$ be a model for?

39–42 ▪ Sketch each vector field and describe its graph.

39. $\mathbf{F}(x, y) = \mathbf{i} + 2\mathbf{j}$

40. $\mathbf{F}(x, y) = x\mathbf{i} + y\mathbf{j}$

41. $\mathbf{F}(x, y) = (x\mathbf{i} + y\mathbf{j})/\sqrt{x^2 + y^2}$

42. $\mathbf{F}(x, y) = x\mathbf{i}$

43. Match each vector field with its graph: $\mathbf{F} = y\mathbf{i}$, $\mathbf{G} = -x\mathbf{i} - y\mathbf{j}$, and $\mathbf{H} = \sin x\mathbf{i} + \sin y\mathbf{j}$.

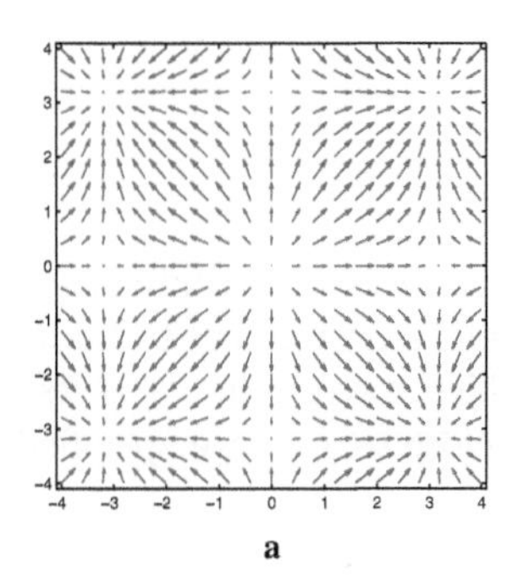

a

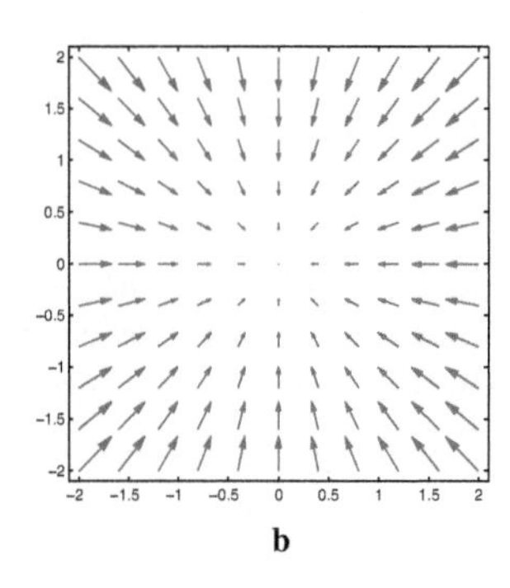

b

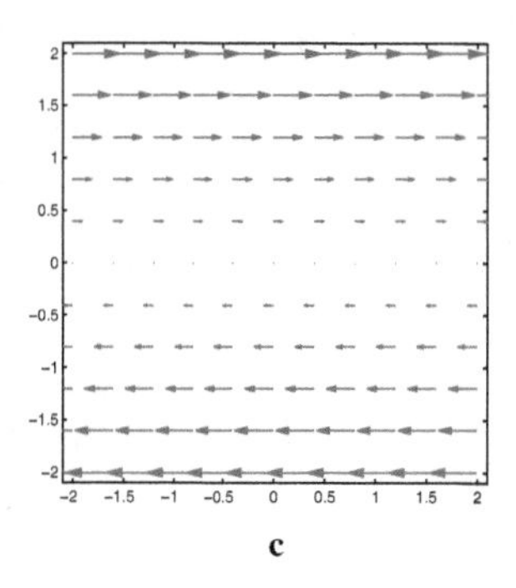

c

44. Prove that the level curves of the linear function $l(x, y) = mx + ny + p$, where $m \neq 0$ or $n \neq 0$, are parallel lines. By looking at the y-intercepts, show that the distance between the lines corresponding to equal difference values is the same (i.e., the lines are equally spaced).

3 Limits and Continuity

In this section we discuss the basic ideas of calculus: **limits** and **continuity**. A rigorous treatment of limits requires more solid background preparation and a lot more space than we are ready to spend in this book. We present the material on a more intuitive level, leaving the hard stuff to more advanced calculus and analysis courses.

Limits

Recall that we say that

$$\lim_{x \to a} f(x) = L$$

if we can make the values $f(x)$ as close to L as desired by taking x close enough to a, but not equal to a. Taking "x close enough to a" reduces to finding an interval around a (usually a symmetric interval, which means that a is its midpoint) so that for every x in that interval, $f(x)$ is as close to L as we specified it to be.

Now we generalize this notion to functions of two variables (an extension to functions of three, four, or more variables is almost identical, except that, as mentioned in Section 2, we have no good way of drawing graphs and providing adequate visual interpretations).

We state the informal definition of the limit, since that will suffice for what we need in this book. Although the definition of the limit for two variables is a straightforward extension of the one-variable case, working with it, as well as using the formal definition and constructing proofs of properties of limits, is more challenging (and usually covered in advanced calculus or analysis courses).

Definition 7 **Limit of a Function of Two Variables**

We say that the *limit of f as (x, y) approaches* (a, b) is equal to L and write

$$\lim_{(x,y) \to (a,b)} f(x, y) = L$$

if we can make the values $f(x, y)$ as close to L as desired by taking (x, y) close enough to (a, b), but not equal to (a, b). ▲

In the one-variable case (where the domain of a function is a subset of the real numbers), we use intervals to describe closeness. For instance, we say $-0.05 < x < 0.05$, or $1.999 < x < 2.001$. The domain of a function of two variables is a subset of the xy-plane, and we need something to replace an interval as a way of measuring closeness.

Definition 8 **Open Disk**

An *open disk* $B_r(a, b)$ of radius r centred at (a, b) is the set of points whose distance from (a, b) is less than r. ▲

The distance between points (x, y) and (a, b) in $\mathbb{R}^2$ is given by

$$d = \sqrt{(x-a)^2 + (y-b)^2}$$

We rewrite the definition of the open disk as

$$B_r(a, b) = \{(x, y) \in \mathbb{R}^2 \mid \sqrt{(x-a)^2 + (y-b)^2} < r\}$$

In this way, finding (x, y) *close enough to* (a, b), as requested by Definition 7, reduces to finding an open disk $B_r(a, b)$ so that whenever (x, y) is in $B_r(a, b)$ the values of $f(x, y)$ lie as close to L as desired.

For this to work, we have to assume that f is defined on some open disk around (a, b). Therefore, in all situations where we use limits (such as continuity or derivatives), we assume that all functions involved have this property.

Example 3.1 Geometric Reasoning about the Limit of a Function

Figure 3.1a shows a contour diagram of a function f that satisfies

$$\lim_{(x,y)\to(2,3)} f(x, y) = 4$$

(We assume that the values of f in the region between the two contour curves lie between the values of f at the two curves.)

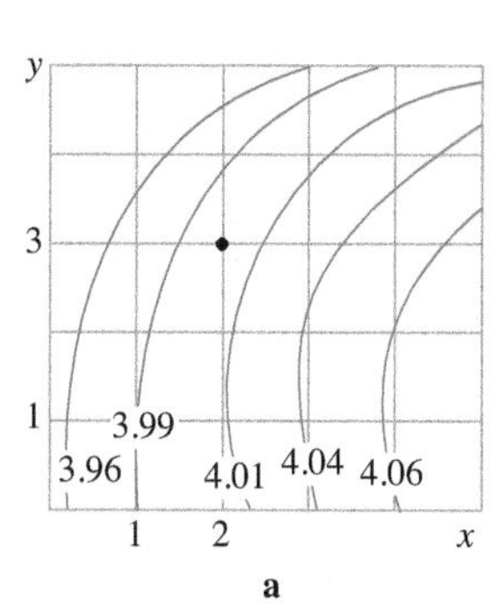

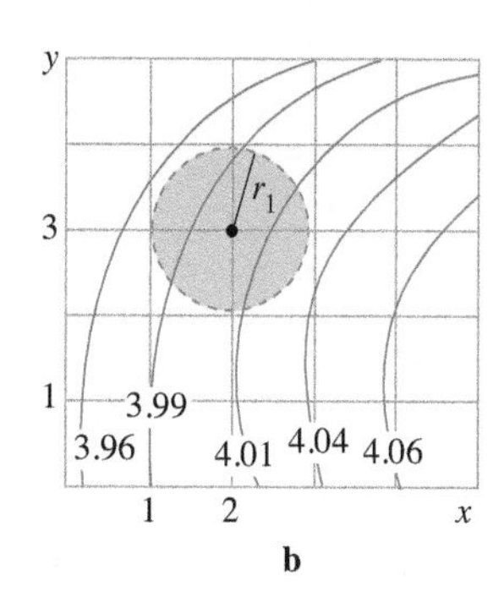

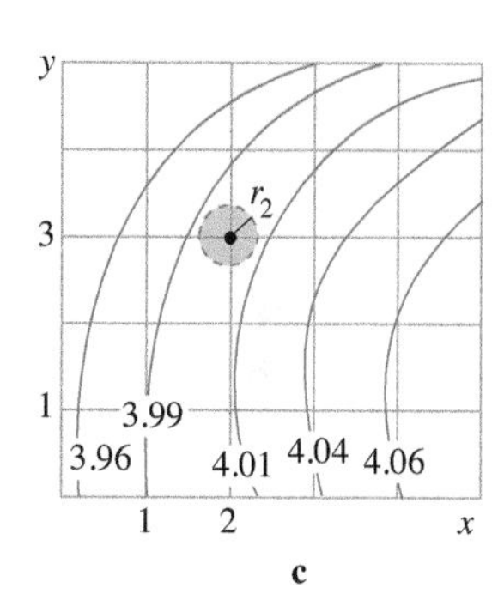

FIGURE 3.1

Contour diagram and open disks around $(2, 3)$

Suppose we wish to make $f(x, y)$ closer to 4 than 0.5, i.e.,

$$3.95 < f(x, y) < 4.05$$

Looking at the contour diagram, we identify the open disk $B_{r_1}(2, 3)$; see Figure 3.1b. For any point (x, y) in it, the values $f(x, y)$ lie between 3.96 and 4.04, which is a bit better than what we need. To make $f(x, y)$ even closer to 4, say

$$3.99 < f(x, y) < 4.01$$

we need to pick a smaller open disk—we call it $B_{r_2}(2, 3)$; see Figure 3.1c. For every point (x, y) whose distance from $(2, 3)$ is less than r_2, the values of f fall within the desired bounds of 3.99 and 4.01.

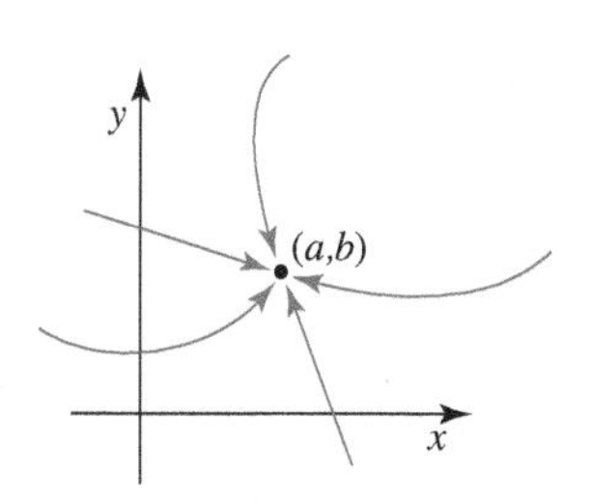

FIGURE 3.2

Paths leading to (a, b)

Recall that in the one-variable case we can determine the limit of a function $f(x)$ as $x \to a$ by investigating one-sided limits. If the left limit and the right limit are equal to the same real number L, then the limit of $f(x)$ as $x \to a$ exists and is equal to L. Otherwise, the limit of $f(x)$ as $x \to a$ does not exist.

We cannot use the same approach for functions of two variables—there are infinitely many ways (we call them paths) in which one can approach (a, b); see Figure 3.2.

To prove that $\lim_{(x,y)\to(a,b)} f(x, y) = L$ we would have to examine *all* paths leading to (a, b) and show that $f(x, y)$ approaches L along every single one of them. As this is not possible, we need other ways of proving that $\lim_{(x,y)\to(a,b)} f(x, y) = L$. (We do not discuss these strategies in this book.)

There is good news here as well: if we determine that $f(x, y)$ approaches L_1 as $(x, y) \to (a, b)$ along one path and $f(x, y)$ approaches L_2 as $(x, y) \to (a, b)$ along another, and if $L_1 \neq L_2$, then $\lim_{(x,y)\to(a,b)} f(x, y)$ does not exist. (Of course, if f approaches ∞ or $-\infty$ along any path, then $\lim_{(x,y)\to(a,b)} f(x, y)$ does not exist.)

Example 3.2 Numerical Investigation of a Limit

We use numerical means to investigate the limits of

$$f(x,y) = \frac{x+3}{xy+2} \quad \text{and} \quad g(x,y) = \frac{xy}{x^2+y^2}$$

when $(x,y) \to (0,0)$. As in the one-variable case, we build a table of values. But in this case, besides the variables (x,y) we also record the distance from (x,y) to $(0,0)$, to make sure that we pick points that are getting close to $(0,0)$.

Table 3.1

(x,y)	Distance to $(0,0)$	$f(x,y)$	$g(x,y)$
$(1, 0.9)$	1.345362405	1.37931034	0.4972375691
$(0.5, 0.4)$	0.6403124237	1.59090909	0.4878048780
$(0.01, 0.015)$	0.01802775638	1.504887133	0.4615384615
$(0.005, 0.0045)$	0.006726812024	1.502483097	0.4972375691
$(0.0001, 0.0001)$	0.0001414213562	1.500049992	0.5

Looking at the values of $f(x,y)$, we claim that

$$\lim_{(x,y)\to(0,0)} f(x,y) = 1.5 \quad \text{and} \quad \lim_{(x,y)\to(0,0)} g(x,y) = 0.5$$

To get a better feel for the limits, we take more values for (x,y):

Table 3.2

(x,y)	Distance to $(0,0)$	$f(x,y)$	$g(x,y)$
$(1, 1.9)$	2.147091055	1.025641026	0.4121475054
$(0.5, 0.95)$	1.073545528	1.414141414	0.4121475054
$(0.01, 0.019)$	0.02147091055	1.504857039	0.4121475054
$(0.005, 0.0099)$	0.01109098733	1.502462814	0.4024063084
$(0.0001, 0.0002)$	0.0002236067977	1.500049985	0.4

This time, it looks like

$$\lim_{(x,y)\to(0,0)} f(x,y) = 1.5 \quad \text{and} \quad \lim_{(x,y)\to(0,0)} g(x,y) = 0.4$$

It seems that $f(x,y)$ approaches 1.5; however, $g(x,y)$ approaches different values, dependimg on how we pick points that are close to $(0,0)$.

We claim that

$$\lim_{(x,y)\to(0,0)} f(x,y) = 1.5$$

and

$$\lim_{(x,y)\to(0,0)} g(x,y) \text{ does not exist}$$

Later in this section, using limit laws, we will prove that $\lim_{(x,y)\to(0,0)} f(x,y) = 1.5$. First, we show that $\lim_{(x,y)\to(0,0)} g(x,y)$ does not exist.

As mentioned earlier, all we need to do is to show that two different paths approaching $(0,0)$ lead to different values for the limit.

Let (x, y) approach $(0, 0)$ along $y = x$. In that case,

$$\lim_{(x,y)\to(0,0)} \frac{xy}{x^2 + y^2} = \lim_{x\to 0} \frac{x \cdot x}{x^2 + x^2} = \lim_{x\to 0} \frac{x^2}{2x^2} = \lim_{x\to 0} \frac{1}{2} = \frac{1}{2} = 0.5$$

Note that the points in Table 3.1 were chosen so that they lie on the line $y = x$, or near it. Now let (x, y) approach $(0, 0)$ along the line $y = 2x$ (Figure 3.3):

$$\lim_{(x,y)\to(0,0)} \frac{xy}{x^2 + y^2} = \lim_{x\to 0} \frac{x \cdot 2x}{x^2 + (2x)^2} = \lim_{x\to 0} \frac{2x^2}{5x^2} = \lim_{x\to 0} \frac{2}{5} = \frac{2}{5} = 0.4$$

This explains the values in Table 3.2, where the points lie on, or near, the line $y = 2x$. Looking at the above, we conclude that $\lim_{(x,y)\to(0,0)} g(x, y)$ does not exist. See Figure 2.6 in Section 2 for the graph of g.

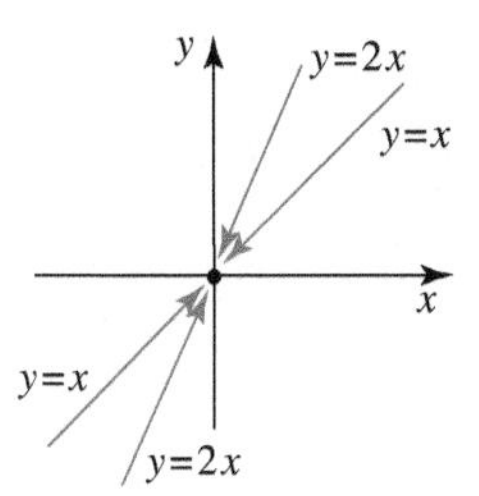

FIGURE 3.3

Paths $y = x$ and $y = 2x$ leading to $(0, 0)$

Example 3.3 Proving That a Limit Does Not Exist

Show that $\lim_{(x,y)\to(0,0)} f(x, y)$ does not exist, where

$$f(x, y) = \frac{y^2 - x^2}{2x^2 + 3y^2}$$

To prove this, we need to find two paths to $(0, 0)$ along which the values for the limit differ. If we let (x, y) approach $(0, 0)$ along the x-axis (i.e., when $y = 0$), then

$$\lim_{(x,y)\to(0,0)} \frac{y^2 - x^2}{2x^2 + 3y^2} = \lim_{x\to 0} \frac{-x^2}{2x^2} = -\frac{1}{2}$$

However, if (x, y) approaches $(0, 0)$ along the y-axis (i.e., when $x = 0$), then

$$\lim_{(x,y)\to(0,0)} \frac{y^2 - x^2}{2x^2 + 3y^2} = \lim_{x\to 0} \frac{y^2}{3y^2} = \frac{1}{3}$$

Thus, the limit of f at $(0, 0)$ does not exist. Note that this is hard (probably impossible) to detect from the graph; see Figure 3.4.

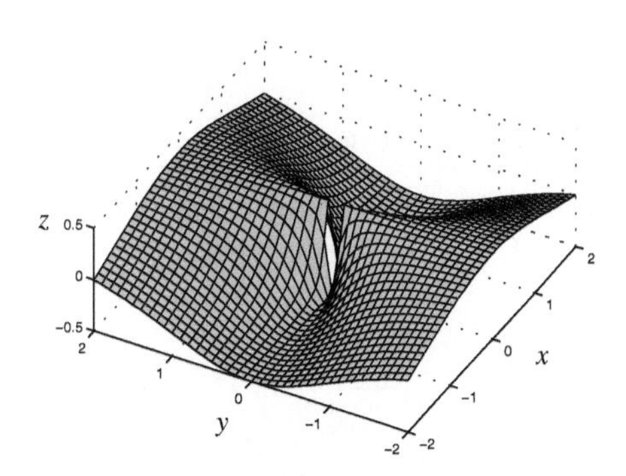

FIGURE 3.4

The graph of the function $f(x, y) = \dfrac{y^2 - x^2}{2x^2 + 3y^2}$

In general, when proving that a limit does not exist, it is not as easy to find paths along which the values of the limit differ. Here, we focused on approaches using straight lines; see Exercises 32 and 33, where we use approaches along curves.

Next, we state (without proof) the limit laws for functions of two variables. In order to prove these properties, we would have to formally define the limit (this is usually done in advanced calculus courses).

Theorem 1 Limit Laws

Assume that $\lim_{(x,y)\to(a,b)} f(x,y)$ and $\lim_{(x,y)\to(a,b)} g(x,y)$ exist (i.e., are real numbers). Then

(a) $\lim_{(x,y)\to(a,b)} (f(x,y)+g(x,y))$ and $\lim_{(x,y)\to(a,b)} (f(x,y)-g(x,y))$ exist, and

$$\lim_{(x,y)\to(a,b)} (f(x,y)\pm g(x,y)) = \lim_{(x,y)\to(a,b)} f(x,y) \pm \lim_{(x,y)\to(a,b)} g(x,y)$$

(b) For any constant c, $\lim_{(x,y)\to(a,b)} (cf(x,y))$ exists, and

$$\lim_{(x,y)\to(a,b)} (cf(x,y)) = c \lim_{(x,y)\to(a,b)} f(x,y)$$

(c) $\lim_{(x,y)\to(a,b)} (f(x,y)g(x,y))$ exists, and

$$\lim_{(x,y)\to(a,b)} (f(x,y)g(x,y)) = \left(\lim_{(x,y)\to(a,b)} f(x,y)\right)\left(\lim_{(x,y)\to(a,b)} g(x,y)\right)$$

(d) If $\lim_{(x,y)\to(a,b)} g(x,y) \neq 0$, then $\lim_{(x,y)\to(a,b)} \dfrac{f(x,y)}{g(x,y)}$ exists, and

$$\lim_{(x,y)\to(a,b)} \frac{f(x,y)}{g(x,y)} = \frac{\lim_{(x,y)\to(a,b)} f(x,y)}{\lim_{(x,y)\to(a,b)} g(x,y)}$$

The limit laws, together with the following basic rules, will allow us to calculate the limits of polynomials and rational functions.

For the function $f(x,y)=x$,

$$\lim_{(x,y)\to(a,b)} f(x,y) = \lim_{(x,y)\to(a,b)} x = a \tag{3.1}$$

Likewise, for $f(x,y)=y$,

$$\lim_{(x,y)\to(a,b)} y = b \tag{3.2}$$

For a constant function $f(x,y)=c$,

$$\lim_{(x,y)\to(a,b)} c = c \tag{3.3}$$

Example 3.4 Using the Limit Laws

Compute $\lim_{(x,y)\to(-1,3)} (x^2y - 7 + y^3)$.

Using the limit laws, we write

$$\lim_{(x,y)\to(-1,3)} (x^2y-7+y^3) = \lim_{(x,y)\to(-1,3)} x^2y - \lim_{(x,y)\to(-1,3)} 7 + \lim_{(x,y)\to(-1,3)} y^3$$

$$= \left(\lim_{(x,y)\to(-1,3)} x\right)\left(\lim_{(x,y)\to(-1,3)} x\right)\left(\lim_{(x,y)\to(-1,3)} y\right) - \lim_{(x,y)\to(-1,3)} 7$$

$$+ \left(\lim_{(x,y)\to(-1,3)} y\right)\left(\lim_{(x,y)\to(-1,3)} y\right)\left(\lim_{(x,y)\to(-1,3)} y\right)$$

$$= (-1)(-1)(3) - 7 + (3)(3)(3) = 3 - 7 + 27 = 23.$$

As in the case of functions of one variable, we can prove the following statement.

Theorem 2 **Direct Substitution**

If $f(x, y)$ is a polynomial or a rational function (in which case (a, b) must be in the domain of f), then

$$\lim_{(x,y)\to(a,b)} f(x, y) = f(a, b)$$

Thus,

$$\lim_{(x,y)\to(4,0)} \frac{3x^2 - yx^4}{x - y^2 - 2} = \frac{3(4)^2 - (0)(4)^4}{4 - (0)^2 - 2} = \frac{48}{2} = 24$$

and (for the function $f(x, y)$ from Example 3.2)

$$\lim_{(x,y)\to(0,0)} \frac{x + 3}{xy + 2} = \frac{0 + 3}{(0)(0) + 2} = \frac{3}{2}$$

Continuity

Generalizing the definition for functions of one variable, we obtain the following.

Definition 9 **Continuity for a Function of Two Variables**

A function f is *continuous* at a point (a, b) in $\mathbb{R}^2$ if

(a) $\lim_{(x,y)\to(a,b)} f(x, y)$ exists (i.e., is a real number)

(b) $f(x, y)$ is defined at (a, b)

(c) $\lim_{(x,y)\to(a,b)} f(x, y) = f(a, b)$

The requirement (c) states that the values $f(x, y)$ get closer and closer to $f(a, b)$ as (x, y) gets closer and closer to (a, b). As a matter of fact, we can make the values of $f(x, y)$ as close to $f(a, b)$ as we wish by taking (x, y) close enough to (a, b). Thus, small changes in the independent variables produce small changes in the values of a continuous function. Understood intuitively, the graph of a continuous function $f(x, y)$ cannot have holes, gaps, jumps, or tears.

Using Definition 9, we can prove various properties of continuous functions.

Theorem 3 **Properties of Continuous Functions**

Assume that the functions $f(x, y)$ and $g(x, y)$ are continuous at (a, b). Then we have the following:

(a) The sum $f(x, y) + g(x, y)$ and the difference $f(x, y) - g(x, y)$ are continuous at (a, b).

(b) The products $cf(x, y)$ (c is a real number) and $f(x, y)g(x, y)$ are continuous at (a, b).

(c) The quotient $f(x, y)/g(x, y)$ is continuous at (a, b) if $g(a, b) \neq 0$.

Statements (a) and (b) in the next theorem are consequences of formulas (3.1) to (3.3) listed after Theorem 1. The remaining two statements follow from Theorem 3.

Theorem 4 **Basic Continuous Functions**

(a) The constant function $f(x, y) = c$ (c is a real number) is continuous at all (a, b) in $\mathbb{R}^2$.

(b) The functions $f(x, y) = x$ and $f(x, y) = y$ are continuous at all (a, b) in $\mathbb{R}^2$.

(c) A polynomial in two variables is continuous at all (a, b) in $\mathbb{R}^2$.

(d) A rational function $f(x, y)/g(x, y)$ is continuous at all points (a, b) in $\mathbb{R}^2$ where $g(a, b) \neq 0$.

Next, we look into the composition of functions.

Consider an example: the function

$$h(x, y) = \sin(x^3 - 2xy)$$

can be wriiten as a composition $f \circ g$ of the function $g(x, y) = x^3 - 2xy$ of two variables and the function $f(t) = \sin t$ of one variable. If $f(t) = 2\sqrt{t} - 1/t$ and $g(x, y) = x^2 + y^2 - 1$, then

$$(f \circ g)(x, y) = f(g(x, y)) = f(x^2 + y^2 - 1) = 2\sqrt{x^2 + y^2 - 1} - \frac{1}{x^2 + y^2 - 1}$$

Note that the inner function in the composition is a function of two variables, whereas the outer is a function of one variable.

Theorem 5 Continuity of the Composition of Functions

Assume that $g(x, y)$ is continuous at (a, b) in $\mathbb{R}^2$ and $f(t)$ is continuous at $g(a, b)$. Then the composition

$$(f \circ g)(x, y) = f(g(x, y))$$

is continuous at (a, b).

Thus, to our list of continuous functions of two variables we add all compositions involving continuous functions of one variable (exponential, logarithmic, absolute value, trigonometric, and inverse trigonometric functions).

Example 3.5 Continuous Functions

Since $f(x, y) = x^3 - 2xy$ is a polynomial, it is continuous at all (a, b) in $\mathbb{R}^2$. The function $g(t) = \sin t$ is continuous for all real numbers, and so Theorem 5 implies that $h(x, y) = \sin(x^3 - 2xy)$ is continuous at all points (a, b) in $\mathbb{R}^2$.

The function $f(t) = 2\sqrt{t} - 1/t$ is continuous whenever $t > 0$. Since the polynomial $g(x, y) = x^2 + y^2 - 1$ is continuous at all (x, y), we conclude that the composition

$$(f \circ g)(x, y) = 2\sqrt{x^2 + y^2 - 1} - \frac{1}{x^2 + y^2 - 1}$$

is continuous at all (x, y) such that $x^2 + y^2 - 1 > 0$, i.e., $x^2 + y^2 > 1$. In words, $f \circ g$ is continuous at all points in the xy-plane that lie outside the circle of radius 1 centred at the origin.

Example 3.6 Functions with Discontinuities

In the text following Example 2.5 in Section 2 we mentioned that the computer-generated plots of $g(x, y) = xy(x^2 + y^2)^{-1}$ and $h(x, y) = \arctan(y/x)$ fail to show important properties of these functions, or are incorrect (see Figure 2.6).

Earlier in this section we showed that

$$\lim_{(x,y)\to(0,0)} \frac{xy}{x^2 + y^2}$$

does not exist, which implies that $g(x, y) = xy(x^2 + y^2)^{-1}$ is not continuous at $(0, 0)$. This is something we cannot see in the computer-generated graph.

number of rabbits is a decreasing function of handling time (Figure 4.1a). In other words, the longer the handling time, the fewer rabbits are captured.

Now fix the handling time, say $T_h = 0.3$ (in certain units, for instance weeks). The number of rabbits captured is now a function of density:

$$c(N, 0.3) = \frac{N}{1 + 0.3N}$$

From

$$\frac{d}{dN}c(N, 0.3) = \frac{(1 + 0.3N) - N(0.3)}{(1 + 0.3N)^2} = \frac{1}{(1 + 0.3N)^2} > 0$$

we conclude that the number of rabbits captured increases with density (Figure 4.1b). This makes sense: the more rabbits there are (in some region), keeping the handling time fixed, the more rabbits will be captured.

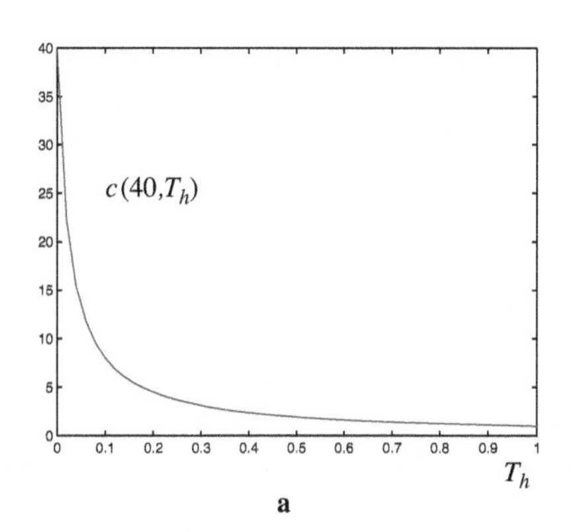

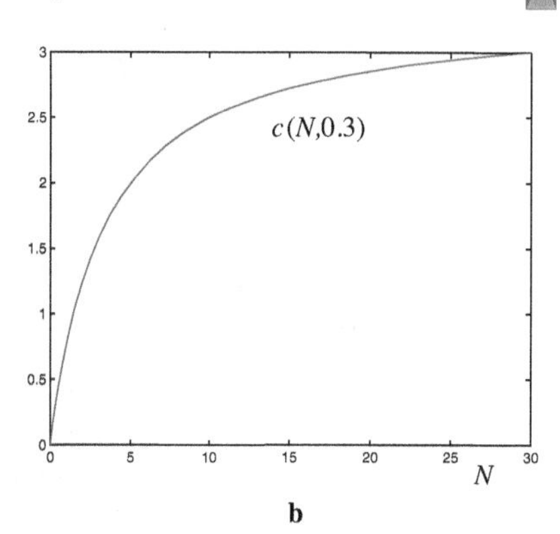

FIGURE 4.1

The number of rabbits captured as a function of handling time and density

The derivatives that we calculated in the previous example are called *partial derivatives.* Intuitively, for a function $f(x, y)$ of two variables, the *partial derivative with respect to* x is the derivative of f viewed as a function of x only (i.e., y is taken to be constant). Likewise, the *partial derivative of* f *with respect to* y is the derivative of f viewed as a function of y only (in other words, x is a parameter, i.e., assumed constant).

How do we find the change in the number of rabbits captured if *both* T_h and N change at the same time? We will answer this question in Section 5. In this section, we focus on studying how a function of several variables changes when *only one* of its variables changes at a time. As usual, we develop concepts and work with functions of two variables, but keep in mind that what we say holds for a function of any number of variables.

Recall that the derivative of a function $y = f(x)$ is given by

$$f'(x) = \lim_{h \to 0} \frac{f(x + h) - f(x)}{h}$$

provided that the limit exists (i.e., is a real number).

Definition 10 **Partial Derivatives**

Let $f(x, y)$ be a real-valued function of two variables x and y. The *partial derivative with respect to* x is the real-valued function $\partial f/\partial x$ defined by

$$\frac{\partial f}{\partial x}(x, y) = \lim_{h \to 0} \frac{f(x + h, y) - f(x, y)}{h}$$

provided that the limit exists. The *partial derivative with respect to* y is the real-valued function $\partial f/\partial y$ defined by

$$\frac{\partial f}{\partial y}(x, y) = \lim_{h \to 0} \frac{f(x, y + h) - f(x, y)}{h}$$

provided that the limit exists.

Thus, $\partial f/\partial x$ (or $\partial f/\partial y$) can be obtained by regarding y (or x) as constant and applying standard rules for differentiation of functions of one variable.

To denote the partial derivatives of $z = f(x, y)$, we can use any of

$$\frac{\partial f}{\partial x} = D_1 f = f_x = \frac{\partial z}{\partial x} = z_x \quad \text{and} \quad \frac{\partial f}{\partial y} = D_2 f = f_y = \frac{\partial z}{\partial y} = z_y$$

If we wish to keep track of the variables, then we write

$$\frac{\partial f}{\partial x}(x, y) = D_1 f(x, y) = f_x(x, y) = \frac{\partial z}{\partial x}(x, y) = z_x(x, y)$$

and, similarly, for the partial derivative with respect to y.

Example 4.2 Calculating Partial Derivatives

Compute $\partial f/\partial x$ and $\partial f/\partial y$ for $f(x, y) = x^3 \sin(x^2 + e^y)$.

Regarding y as constant and using the product rule and the chain rule, we find

$$\begin{aligned}\frac{\partial f}{\partial x}(x, y) &= 3x^2 \sin(x^2 + e^y) + x^3 \cos(x^2 + e^y) \cdot 2x \\ &= 3x^2 \sin(x^2 + e^y) + 2x^4 \cos(x^2 + e^y)\end{aligned}$$

Keeping x fixed,

$$\frac{\partial f}{\partial y}(x, y) = x^3 \cos(x^2 + e^y) \cdot e^y = x^3 e^y \cos(x^2 + e^y)$$

In some cases, we need to use the definition to calculate partial derivatives.

Example 4.3 Using the Definition to Calculate Partial Derivatives

Find $\partial f/\partial x(0, 0)$ for the function $f(x, y) = (x^3 + y^3)^{1/3}$.

Using the chain rule while keeping y fixed, we get

$$\frac{\partial f}{\partial x}(x, y) = \frac{1}{3}\left(x^3 + y^3\right)^{-2/3} \cdot 3x^2 = \frac{x^2}{(x^3 + y^3)^{2/3}}$$

This is fine, but it does not help us answer the question: at $(0, 0)$, the denominator is zero. So we use Definition 10 instead:

$$\begin{aligned}\frac{\partial f}{\partial x}(0, 0) &= \lim_{h \to 0} \frac{f(0 + h, 0) - f(0, 0)}{h} \\ &= \lim_{h \to 0} \frac{f(h, 0) - f(0, 0)}{h} \\ &= \lim_{h \to 0} \frac{(h^3)^{1/3} - 0}{h} \\ &= \lim_{h \to 0} \frac{h}{h} = 1\end{aligned}$$

Thus, $\partial f/\partial x(0, 0) = 1$.

Example 4.4 A Smart Way of Calculating Partial Derivatives

Compute $\partial f/\partial y(1, 3)$ for the function $f(x, y) = y^3 e^{(x-1)\sin y} + x^2 \ln y$.

Calculating the derivative of $f(x, y)$ looks complicated. Is there a way we can make it simpler?

We are not asked to calculate the partial derivative $\partial f/\partial y$ at a *general point* (x, y) but at a *specific point* $(1, 3)$. Keeping in mind that we are supposed to keep x fixed, we calculate $f(1, y)$ first, and then differentiate.

By substituting $x = 1$, we get

$$f(1, y) = y^3 e^{(1-1)\sin y} + (1)^2 \ln y = y^3 + \ln y$$

which is lot simpler to differentiate than the original function. Thus,

$$\frac{\partial f}{\partial y}(1, y) = 3y^2 + \frac{1}{y}$$

and $\partial f/\partial y(1, 3) = 3(3)^2 + 1/3 = 82/3$. ▲

Example 4.5 **Interpreting Partial Derivatives**

The function

$$c(x, t) = \frac{1}{\sqrt{4\pi t}} e^{-x^2/4t}$$

(introduced in Example 2.11 in Section 2) describes the concentration of a pollutant at a location x units away from the source at time $t > 0$ (for simplicity, we take the diffusion coefficient D to be 1). Calculate the partial derivative $c_t(2, 1)$ and interpret the result.

▶ We write

$$c(x, t) = \frac{1}{\sqrt{4\pi}} t^{-1/2} e^{(-x^2/4)t^{-1}}$$

and now use the product rule and the chain rule (while keeping x fixed)

$$\begin{aligned} c_t(x, t) &= \frac{1}{\sqrt{4\pi}} \left(-\frac{1}{2}\right) t^{-3/2} e^{-x^2/4t} + \frac{1}{\sqrt{4\pi t}} e^{-x^2/4t} \left(-\frac{x^2}{4}\right)\left(-\frac{1}{t^2}\right) \\ &= -\frac{1}{2\sqrt{\pi}} \left(\frac{1}{2}\right) \frac{1}{t^{3/2}} e^{-x^2/4t} + \frac{1}{2\sqrt{\pi}\, t^{1/2}} e^{-x^2/4t} \left(-\frac{x^2}{4}\right)\left(-\frac{1}{t^2}\right) \\ &= -\frac{1}{4\sqrt{\pi}\, t^{3/2}} e^{-x^2/4t} + \frac{x^2}{8\sqrt{\pi}\, t^{5/2}} e^{-x^2/4t} \end{aligned}$$

(Note that the strategy introduced in Example 4.4 would not have helped much —we'd still have to use the product rule.) We compute

$$\begin{aligned} c_t(2, 1) &= -\frac{1}{4\sqrt{\pi}\,(1)^{3/2}} e^{-(2)^2/4(1)} + \frac{(2)^2}{8\sqrt{\pi}\,(1)^{5/2}} e^{-(2)^2/4(1)} \\ &= -\frac{1}{4e\sqrt{\pi}} + \frac{1}{2e\sqrt{\pi}} \\ &= \frac{1}{4e\sqrt{\pi}} \approx 0.0519 \end{aligned}$$

Thus, at a location $x = 2$ distance units from the source, at time $t = 1$, the pollution increases at a rate of 0.0519 units of concentration per unit of time. ▲

Example 4.6 **Partial Derivatives of a Function of Three Variables**

Find g_x and g_z for the function $g(x, y, z) = e^{yz} \sin(xz)$.

▶ Holding y and z constant (so that e^{yz} is constant), we get

$$g_x(x, y, z) = e^{yz} \cos(xz) \cdot z = z e^{yz} \cos(xz)$$

Keeping x and y constant and using the product rule,

$$\begin{aligned} g_z(x, y, z) &= e^{yz} \cdot y \sin(xz) + e^{yz} \cos(xz) \cdot x \\ &= y\, e^{yz} \sin(xz) + x\, e^{yz} \cos(xz) \end{aligned}$$

▲

Example 4.7 Estimating Partial Derivatives from a Table of Values

Table 4.1 shows the values of the wind chill index $W(T, v)$, with temperatures measured in degrees Celsius and wind speed in kilometres per hour. Estimate $W_T(-15, 30)$, and interpret the result.

Table 4.1

	$T = -25$	$T = -20$	$T = -15$	$T = -10$	$T = -5$
$v = 40$	-40.8	-34.1	-27.4	-20.8	-14.1
$v = 30$	-39.1	-32.6	-26.0	-19.5	-13.0
$v = 20$	-36.8	-30.5	-24.2	-17.9	-11.6

Using the definition of the partial derivative, we write

$$W_T(-15, 30) = \lim_{h \to 0} \frac{W(-15 + h, 30) - W(-15, 30)}{h}$$

What do we do with $W(-15 + h, 30)$? In theory, we let h take on smaller and smaller values so that $-15 + h$ approaches -15. But, by looking at the table, we see that the closest we can get to -15 is either -10 (when $h = 5$) or -20 (when $h = -5$). So, instead of calculating the limit, we approximate:

$$W_T(-15, 30) \approx \frac{W(-15 + h, 30) - W(-15, 30)}{h}$$

With $h = 5$, we get

$$W_T(-15, 30) \approx \frac{W(-10, 30) - W(-15, 30)}{5} = \frac{-19.5 - (-26.0)}{5} = 1.30$$

With $h = -5$, we get

$$W_T(-15, 30) \approx \frac{W(-20, 30) - W(-15, 30)}{-5} = \frac{-32.6 - (-26.0)}{-5} = 1.32$$

Given two estimates, we take the average

$$W_T(-15, 30) \approx \frac{1.30 + 1.32}{2} = 1.31$$

What does this number mean?

When $T = -15^{\circ}$C and $v = 30$ km/h, the wind chill is -26.0. At that moment, the wind chill increases by approximately 1.31 units per degree Celsius increase in temperature, assuming no change in wind speed. For example, based on the fact that $W(-15, 30) = -26.0$, we estimate that the wind chill corresponding to $T = -14^{\circ}$C and $v = 30$ km/h is

$$W(-14, 30) \approx W(-15, 30) + 1.31 = -26.0 + 1.31 = -24.69$$

As well, based on the same value $W(-15, 30) = -26.0$, we estimate that the wind chill $W(-10, 30)$ (corresponding to an increase in temperature of 5°C, assuming that the wind blows at the same speed)

$$W(-10, 30) \approx W(-15, 30) + 1.31 \cdot 5 = -26.0 + 6.55 = -19.45$$

This is a fairly good estimate for the value $W(-10, 30) = 19.5$.

Example 4.8 Estimating Partial Derivatives from a Contour Diagram

Figure 4.2 shows a contour diagram of a function $f(x, y)$. Estimate the value $\partial f/\partial x(2, 3)$.

▶ We use the same idea as in Example 4.7: we approximate the limit of difference quotients by the difference quotient that we can deduce from the diagram.

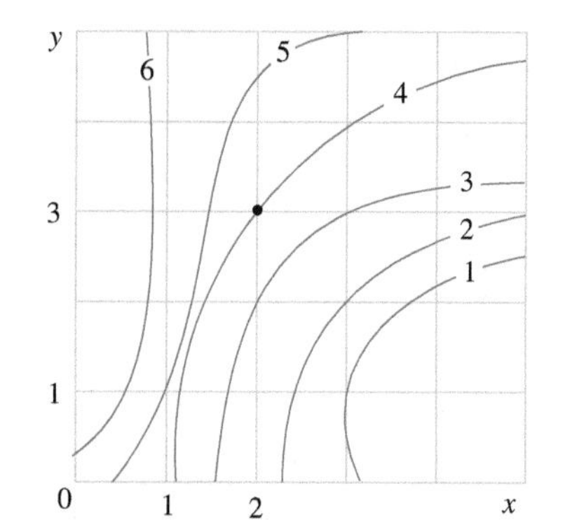

FIGURE 4.2
Contour diagram of $f(x, y)$

Note that the point $(2, 3)$ lies on the contour curve of value 4; thus, $f(2, 3) = 4$. Moving one unit to the right (keeping the y value unchanged), we meet the contour curve of value 3 at the point $(3, 3)$. Thus,

$$\frac{\partial f}{\partial x}(2,3) \approx \frac{f(3,3) - f(2,3)}{3-2} = \frac{3-4}{1} = -1$$

Moving 0.5 units to the left, we arrive at the point $(1.5, 3)$, where f is equal to 5. We estimate

$$\frac{\partial f}{\partial x}(2,3) \approx \frac{f(1.5,3) - f(2,3)}{1.5-2} = \frac{5-4}{-0.5} = -2$$

We take the average of the two for our estimate:

$$\frac{\partial f}{\partial x}(2,3) \approx \frac{-1-2}{2} = -1.5.$$

In a similar way we can estimate the partial derivative with respect to y.

Next, we discuss the geometric interpretation of partial derivatives.

Recall that the graph of a function $z = f(x, y)$ is a surface in $\mathbb{R}^3$. Pick a point (a, b) in the domain of f, and denote by P the corresponding point on the graph (Figure 4.3a). The coordinates of P are $(a, b, f(a, b))$.

Let $y = b$ (so that the y value is fixed). Recall that $y = b$ represents the plane parallel to the xz-plane that crosses the y-axis at $(0, b, 0)$. This plane intersects the graph of f along the curve (call it c_1) whose equation is $g(x) = f(x, b)$. (Algebraically, we combine the equation of the surface $z = f(x, y)$ with the equation of the plane $y = b$.) From

$$g'(a) = \lim_{h\to 0} \frac{g(a+h) - g(a)}{h} = \lim_{h\to 0} \frac{f(a+h,b) - f(a,b)}{h} = \frac{\partial f}{\partial x}(a,b)$$

we conclude that the slope of the tangent to the curve c_1 at P is equal to the partial derivative $\partial f/\partial x(a, b)$.

Likewise, $\partial f/\partial y(a, b)$ is the slope (at P) of the tangent line to the curve c_2, which is the intersection of the surface and the plane $x = a$ (Figure 4.3b).

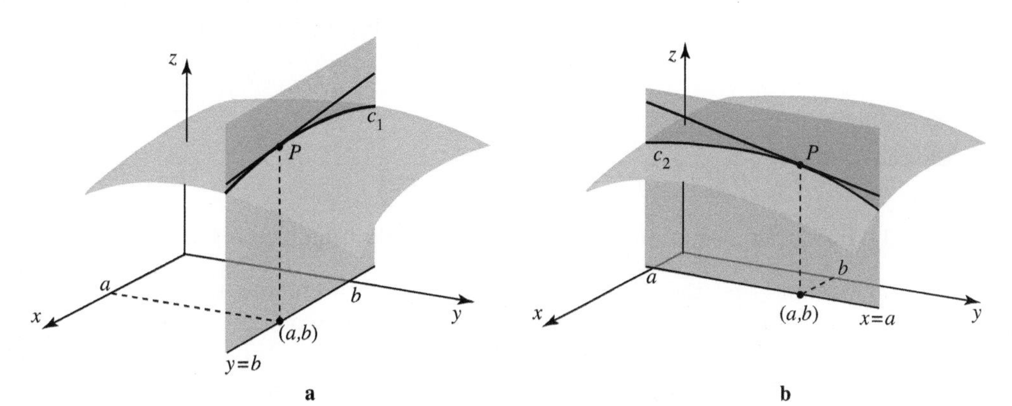

FIGURE 4.3
Interpretation of partial derivatives

Example 4.9 Partial Derivatives as Slopes

Take the function $f(x,y) = \sqrt{16 - x^2 - y^2}$. By interpreting partial derivatives as slopes, find $\partial f/\partial x(1,3)$ and $\partial f/\partial y(1,3)$.

▶ The graph of f is the upper hemisphere (see the text that follows Example 2.7 in Section 2). Intersecting the hemisphere with the plane $y = 3$, we obtain the circle (call it c_1)

$$g(x) = \sqrt{16 - x^2 - (3)^2} = \sqrt{7 - x^2}$$

See Figure 4.4. The slope of the tangent to $g(x)$ at $x = 1$ is equal to the partial derivative $\partial f/\partial x(1,3)$. Thus, from

$$g'(x) = \frac{1}{2}\left(7 - x^2\right)^{-1/2}(-2x) = -\frac{x}{\sqrt{7 - x^2}}$$

we conclude that

$$\frac{\partial f}{\partial x}(1,3) = g'(1) = -\frac{1}{\sqrt{6}}$$

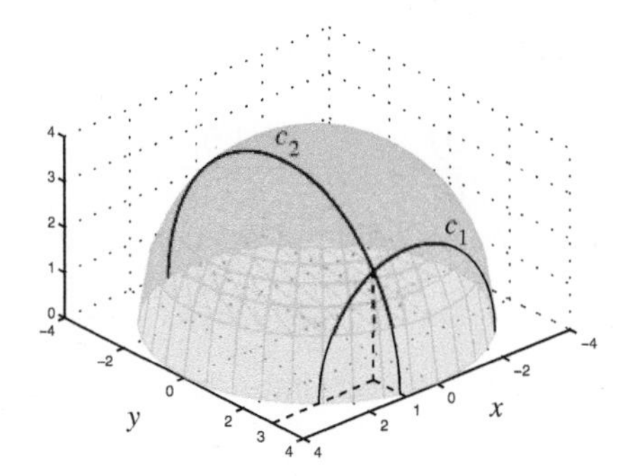

FIGURE 4.4
Partial derivatives as slopes

Likewise, the curve (call it c_2)

$$h(y) = \sqrt{16 - (1)^2 - y^2} = \sqrt{15 - y^2}$$

is the intersection of the hemisphere and the plane $x = 1$. The slope

$$h'(y) = \frac{1}{2}\left(15 - y^2\right)^{-1/2}(-2y) = -\frac{y}{\sqrt{15 - y^2}}$$

at $y = 3$ is equal to the partial derivative $\partial f/\partial y(1,3)$; thus,

$$\frac{\partial f}{\partial y}(1,3) = h'(3) = -\frac{3}{\sqrt{6}}$$

Summary The **partial derivative** of a function of several variables is a way to measure the rate of change of the function as one of its variables changes. To calculate a partial derivative, we fix all but the chosen variable and apply standard differentiation rules. We can estimate partial derivatives from the contour diagrams or from the table of values of a function. As in the one-variable case, a partial derivative can be interpreted as the **slope of the tangent line** to the graph of the function.

4 Exercises

1. Imitating Definition 10, write the limit definition of the partial derivative f_t of a function $f(x,y,t)$.

2. Assume that the function $T(x,y,t)$ models the temperature (in degrees Celsius) at time t in a city located at a longitude of x degrees and a latitude of y degrees. The time t is measured in hours. What is the meaning of the partial derivative $T_t(x,y,t)$? What are its units? What is most likely going to be the sign of $T_y(x,y,t)$ for Winnipeg, Manitoba, in January?

3. Using the contour diagram of Example 4.8, estimate $\partial f/\partial y(2,3)$.

4. Using the contour diagram in Figure 4.2, determine the signs of the partial derivatives $f_x(3,2)$ and $f_y(3,2)$.

5. Sketch a contour diagram of a function for which $\partial f/\partial x > 0$ and $\partial f/\partial y < 0$ at all points (x,y) in $\mathbb{R}^2$.

6. Sketch a contour diagram of a function that satisfies $\partial f/\partial x = 0$ and $\partial f/\partial y > 0$ at all points (x,y) in $\mathbb{R}^2$.

7. Sketch the graph of the surface $z = f(x,y) = x^2$. Explain how to obtain the curve with the property that the slope of its tangent is equal to the partial derivative $f_y(2,1)$. Add the curve and its tangent to your graph of the surface.

8. Sketch the graph of the surface $z = f(x,y) = x^2 + y^2$. Draw the curve with the property that the slope of its tangent is equal to the partial derivative $f_x(2,4)$. Explain your construction.

9–10 ▪ Given is the graph of a function $z = f(x,y)$. Determine the signs of the partial derivatives f_x and f_y at A and B.

9.

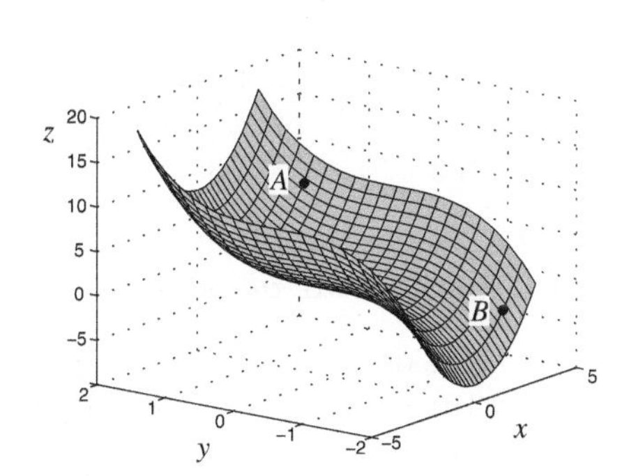

10.

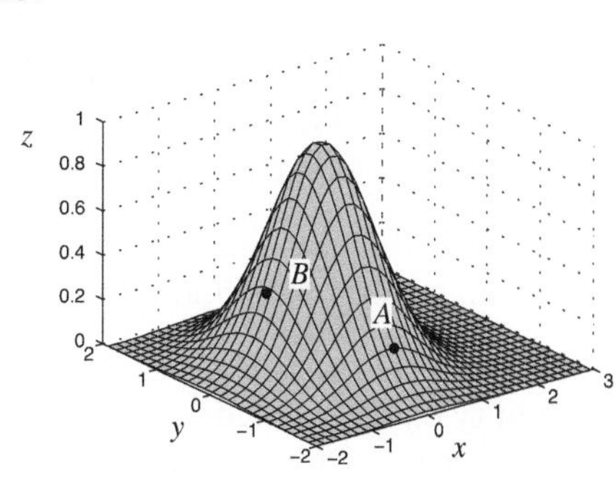

11. Let $f(x,y) = 3 + x^2y^2$. Find the partial derivative $f_x(-2,0)$ and interpret it geometrically.

12. Let $f(x,y) = \sqrt{x^2+y^2}$. Find the partial derivative $f_y(1,1)$ and interpret it geometrically.

13–20 ▪ Find all partial derivatives for each function.

13. $f(x,y) = x^3y^3 - 7xy + 14$

14. $g(x,y) = \dfrac{4x - xy}{x^2 + y^2}$

15. $h(x,t) = \sin(3x - 2t)$

16. $g(x,z) = e^{x-4z^2}$

17. $f(x,t) = \dfrac{1}{4t}e^{-x^2/t}$

18. $f(x,y,z) = \sin(xy) + \cos(xz)$

19. $f(x,w) = w^{-3}\ln(x + w^2)$

20. $g(x,y,w) = xw\sec(xy)$

21. Using the formula for the wind chill (Example 1.5 in Section 1), calculate the value $W_T(-15,30)$ and compare with the estimate obtained in Example 4.7.

22–25 ▪ Find the indicated partial derivative.

22. $g(m,v) = m(1-v^2)^{-1}$; $g_v(60,120)$

23. $f(x,y) = \arctan(3x/y)$; $f_y(1,3)$

24. $h(x,t) = \sqrt{x^2 + xt + 4}$; $h_x(0,4)$

25. $S(m,h) = \sqrt{mh}/6$; $S_h(70,1.6)$

26. Compute $\partial f/\partial y$ for the function $f(x,y) = y^3e^{(x-1)\sin y} + x^2\ln y$ from Example 4.4. Then find $\partial f/\partial y(1,3)$, thus checking the answer provided in the example.

27–31 ▪ A contour diagram of a function f is given below. Answer each question.

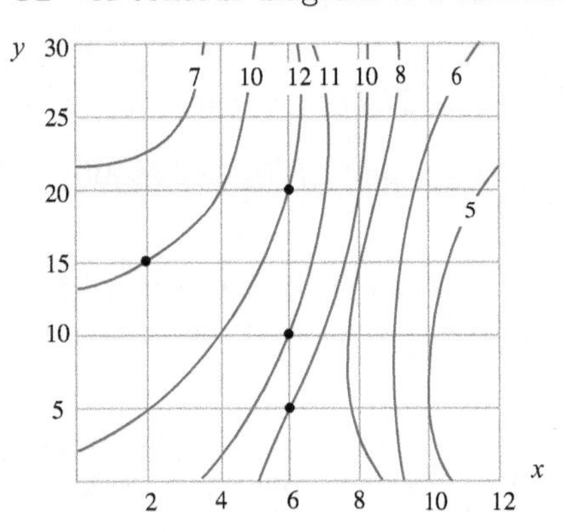

27. Estimate $f_x(6, 10)$.

28. Estimate $f_y(6, 10)$.

29. At $(6, 5)$, which of f_x or f_y is larger?

30. What is the sign of $f_x(6, 20)$?

31. What are the signs of $f_x(2, 15)$ and $f_y(2, 15)$?

32. Recall the body surface area formula $S(m, h) = 0.20247m^{0.425}h^{0.725}$ (Example 1.6 in Section 1; we have dropped the subscript D to simplify the notation), where the height h is in metres and the mass m is in kilograms. Compute and interpret $\partial S/\partial m(50, 1.6)$ and $\partial S/\partial h(50, 1.6)$.

33. Consider the body mass index formula $\text{BMI}(m, h) = m/h^2$ (the height h is in metres and the mass m is in kilograms). Compute and interpret $\partial\text{BMI}/\partial m(60, 1.7)$ and $\partial\text{BMI}/\partial h(60, 1.7)$.

34. The humidex $H(T, h)$ is a measure used by meteorologists to describe the combined effects of heat and humidity on an average person's feeling of hotness. In Table 4.2 we give values of humidex based on measurements of temperature (in degrees Celsius) and relative humidity h (given as a percent). Estimate $H_T(30, 60)$ and interpret your answer.

Table 4.2

	$T = 22$	$T = 26$	$T = 30$	$T = 34$
$h = 70$	27	33	41	49
$h = 60$	25	32	38	46
$h = 50$	24	30	36	43

35. Consider the humidex function $H(T, h)$; see Exercise 34 for details. Based on the values given in Table 4.2, estimate $H_h(26, 60)$ and interpret your answer.

36. Let $u(x, y, t) = e^{-t}\sin(2x)\cos(3y)$ be the vertical displacement of a vibrating membrane from the point (x, y) in the xy-plane at time t. Compute $u_x(\pi/12, \pi/12, 1)$, $u_y(\pi/12, \pi/12, 1)$, and $u_t(\pi/12, \pi/12, 1)$ and give a physical interpretation of your answers.

37. A hiker is standing at the point $(2, 1, 21)$ on a hill whose shape is given by the graph of the function $z = 24 - (x-3)^2 - 2(y-2)^4$. Assume that the x-axis points east and the y-axis points north. In which of the two directions (east or north) is the hill steeper?

38. Consider the type-2 functional response $c(N, T_h) = aN/(1 + aT_hN)$, where $a > 0$ (in Example 4.1 we assumed that $a = 1$).

(a) Find the sign of $\partial c/\partial T_h$ and say what it means in terms of the dependence of c on T_h.

(b) Find $\partial c/\partial N$ and interpret its sign.

(c) Compute the limit of $c(N, T_h)$ as N approaches ∞. Interpret your answer.

5 Tangent Plane, Linearization, and Differentiability

We extend the idea of approximating a function with a **linear function** by defining the **linearization** of a function of two variables. Geometrically, we construct the **tangent plane** to the surface that represents the function. Investigation of the conditions under which a function has a meaningful tangent plane approximation leads to the concept of **differentiability.** At the end, we discuss the **differential** as a way of describing how a function reacts to small changes in all its variables.

Introduction

The graph of a function $y = f(x)$ of one variable is a curve. If f is differentiable at a (i.e., if $f'(a)$ exists), then the equation

$$y - f(a) = f'(a)(x - a)$$

represents the tangent line to the graph of f at the point $(a, f(a))$. This linear function, written in the form

$$L_a(x) = f(a) + f'(a)(x - a)$$

is called the *linearization* of f at $x = a$. Near $x = a$, the function f can be approximated by its tangent line; i.e.,

$$f(x) \approx f(a) + f'(a)(x - a)$$

This formula is called the *linear approximation* (or the *tangent line approximation*) of f at $x = a$. If we zoom in on the graph of f near $x = a$, it looks more and more straight, resembling its tangent line more and more closely.

Example 5.1 Linearization and Linear Approximation

Consider the function $f(x) = x + 2\sqrt{x}$. From $f'(x) = 1 + 1/\sqrt{x}$, we compute the linearization of f at $a = 1$:

$$L_1(x) = f(1) + f'(1)(x - 1) = 3 + 2(x - 1) = 2x + 1$$

Take a point near $a = 1$, say $x = 1.05$. Then

$$L_1(1.05) = 2(1.05) + 1 = 3.1$$

is an approximation of $f(1.05) = 1.05 + 2\sqrt{1.05} = 3.09939$. For values of x far from $a = 1$, the linearization makes no sense: for instance $L_1(10) = 21$, whereas $f(10) = 10 + 2\sqrt{10} \approx 16.32456$.

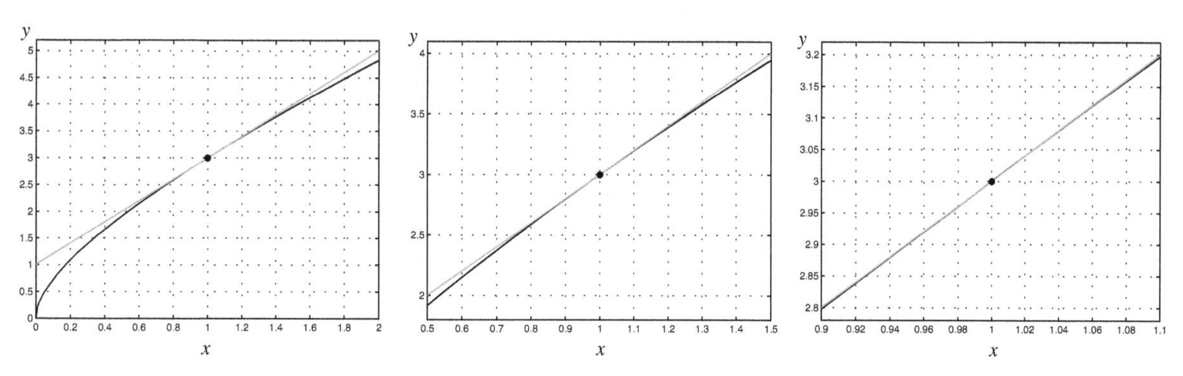

FIGURE 5.1

The function $f(x) = x+2\sqrt{x}$ and its linearization

Figure 5.1 shows both $f(x)$ and $L_1(x)$. As we zoom in on the graphs near $a = 1$, we see that $f(x)$ becomes indistinguishable from its linearization (i.e., its tangent line).

Now we extend this concept—approximating a function with a linear function—to functions of two variables.

Tangent Plane

Take a real-valued function $f(x, y)$ of two variables, and assume that its partial derivatives f_x and f_y are continuous (we will comment on this assumption, and state the reasons why it's needed, later in this section). The graph of $f(x, y)$ is a surface S in $\mathbb{R}^3$.

Pick a point (a, b) in the domain of f, and, as in Section 4, draw the curves that are obtained by intersecting the surface S with the planes $y = b$ (curve c_1) and $x = a$ (curve c_2); see Figure 5.2a. Note that the point $P = (a, b, f(a, b))$ lies on both curves.

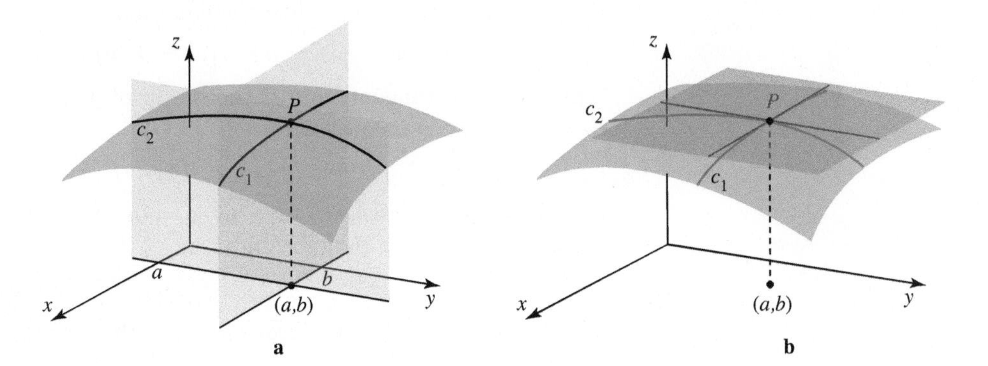

FIGURE 5.2
Surface S and the tangent plane

Definition 11 Tangent Plane
The plane that contains the point P and the tangent lines to the curves c_1 and c_2 at P is called the *tangent plane to the surface* $z = f(x, y)$ *at* P.

See Figure 5.2b. It can be proven that the tangent plane at P contains the tangent line (at P) to *any* curve that lies on the surface S and goes through P. Each of these curves is closely approximated by its tangent line; thus, the tangent plane closely approximates the surface S near the point of tangency. It is the best linear approximation of the surface near the point of tangency in the sense that will be explained soon.

First, we find the equation of the tangent plane to $z = f(x, y)$ at (a, b). Starting with the general equation $z = Mx + Ny + Q$, we find M, N, and Q.

Since the tangent plane contains the point $(a, b, f(a, b))$, we get

$$f(a, b) = Ma + Nb + Q$$

and $Q = f(a, b) - Ma - Nb$. It follows that

$$\begin{aligned} z &= Mx + Ny + Q \\ &= Mx + Ny + f(a, b) - Ma - Nb \\ &= f(a, b) + M(x - a) + N(y - b) \end{aligned}$$

The intersection of the tangent plane and the plane $y = b$ is the line (the tangent to c_1 at P) given by

$$z = f(a, b) + M(x - a)$$

From Section 4 we know that the slope of the tangent at P is equal to the partial derivative of f at P; thus (since we kept y fixed), $M = f_x(a, b)$. Using the same argument we show that $N = f_y(a, b)$, so that

$$z = f(a, b) + f_x(a, b)(x - a) + f_y(a, b)(y - b)$$

To summarize: if f has continuous partial derivatives at (a, b), then

$$z = f(a, b) + f_x(a, b)(x - a) + f_y(a, b)(y - b)$$

is the equation of the tangent plane to the graph of f at (a, b).

As we zoom in on the surface $z = f(x, y)$ near (a, b), it becomes more and more flat—it looks more and more like its tangent plane. Let's look at an example.

Example 5.2 Tangent Plane

Find the equation of the plane tangent to the surface $z = f(x, y) = x^2 + y^2 + 1$ at $(2, 1)$.

▶ The partial derivatives $f_x = 2x$ and $f_y = 2y$ are polynomials, and hence continuous; thus, we can proceed with our calculation of the tangent plane. From $f(2, 1) = 6$, $f_x(2, 1) = 4$, and $f_y(2, 1) = 2$, we get

$$\begin{aligned} z &= f(2, 1) + f_x(2, 1)(x - 2) + f_y(2, 1)(y - 1) \\ &= 6 + 4(x - 2) + 2(y - 1) \end{aligned}$$

i.e., $z = 4x + 2y - 4$. In Figure 5.3 we zoom in on the surface $z = x^2 + y^2 + 1$ and its tangent plane near $(2, 1)$.

As we zoom in on the contour curves of $z = x^2 + y^2 + 1$ near $(2, 1)$, we see that they resemble, more and more closely, the level curves of the linear function $z = 4x + 2y - 4$ (recall that the contour diagram of a linear function consists of equally spaced parallel lines); see Figure 5.4. ▲

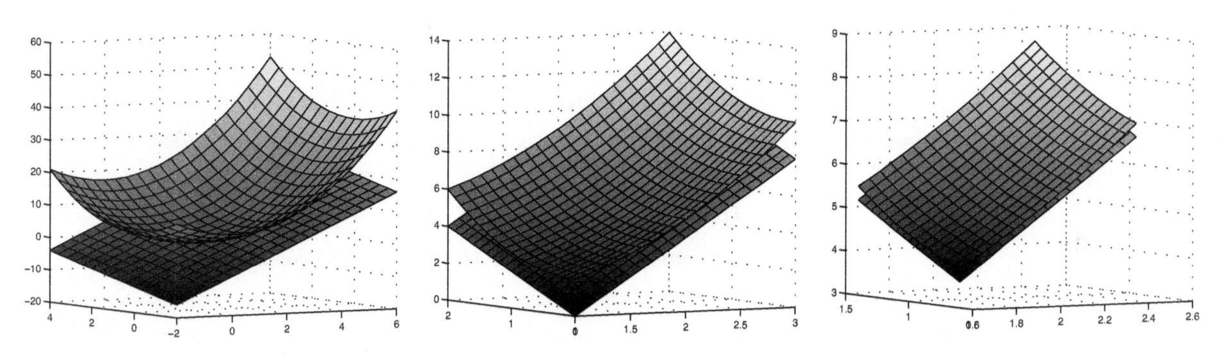

FIGURE 5.3
The surface $f(x, y) = x^2 + y^2 + 1$ and its tangent plane

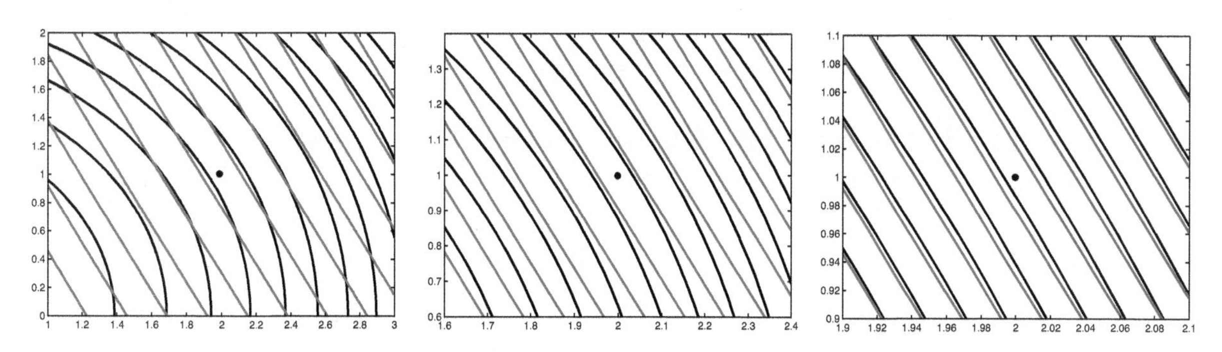

FIGURE 5.4
Zooming in on the contour curves of $f(x, y) = x^2 + y^2 + 1$

Linear Approximation

The tangent plane to the surface $f(x,y) = x^2 + y^2 + 1$ at $(2,1)$ is given by $z = 4x + 2y - 4$ (see Example 5.2). We view z as a function of two variables and denote it by $L(x,y) = 4x + 2y - 4$ or by

$$L_{(2,1)}(x,y) = 4x + 2y - 4$$

when we wish to identify the point of tangency. The function $L_{(2,1)}(x,y)$ is called the *linearization of* $f(x,y)$ *at* $(2,1)$, and the approximation

$$f(x,y) \approx L_{(2,1)}(x,y) = 4x + 2y - 4$$

is called the *linear approximation* (or the *tangent plane approximation*) *of* $f(x,y)$ *at* $(2,1)$. For example,

$$L_{(2,1)}(2.02, 0.95) = 4(2.02) + 2(0.95) - 4 = 5.98$$

approximates the value $f(2.02, 0.95) = 5.9829$. As well,

$$f(1.999, 1.003) \approx L_{(2,1)}(1.999, 1.003),$$

where

$$L_{(2,1)}(1.999, 1.003) = 4(1.999) + 2(1.003) - 4 = 6.002.$$

This is a good approximation of the actual value $f(1.999, 1.003) = 6.00201$.

Definition 12 Linearization and Linear Approximation

Assume that $z = f(x,y)$ has continuous partial derivatives at (a,b). The linear function

$$L_{(a,b)}(x,y) = f(a,b) + f_x(a,b)(x-a) + f_y(a,b)(y-b)$$

is called the *linearization of* f *at* (a,b). The approximation $f(x,y) \approx L_{(a,b)}(x,y)$, i.e.,

$$f(x,y) \approx f(a,b) + f_x(a,b)(x-a) + f_y(a,b)(y-b)$$

is called the *linear approximation* (or the *tangent plane approximation*) *of* f *at* (a,b).

For all we know, the linear approximation as we just defined it is just *one* linear approximation of a function. We will explain in what sense it is the *best linear approximation.* But before we do that, we discuss the reasons why we used the assumption of the continuity of partial derivatives in the construction and in the definiton of the tangent plane.

Look at the following example.

Example 5.3 A Function for Which Linearization Does Not Make Sense

Define

$$g(x,y) = \begin{cases} \dfrac{xy}{x^2 + y^2} & \text{if } (x,y) \neq (0,0) \\ 0 & \text{if } (x,y) = (0,0) \end{cases}$$

The domain of g consists of all points (x,y) in $\mathbb{R}^2$. In Section 3 (see Example 3.2 and the text following it) we showed that

$$\lim_{(x,y)\to(0,0)} g(x,y) \text{ does not exist}$$

Consequently, g is not continuous at $(0,0)$.

Using the definition of the partial derivative, we compute

$$\frac{\partial g}{\partial x}(0,0) = \lim_{h\to 0}\frac{g(0+h,0)-g(0,0)}{h} = \lim_{h\to 0}\frac{\frac{(h)(0)}{h^2+0^2}-0}{h} = \lim_{h\to 0} 0 = 0$$

In the same way, we show that $\partial g/\partial y(0,0) = 0$.

We ignore the assumption on the continuity of partial derivatives in Definition 12 and try to use the formula for the linearization anyway. Since $g(0,0) = 0$, we get

$$\begin{aligned} L_{(0,0)}(x,y) &= g(0,0) + \frac{\partial g}{\partial x}(0,0)(x-0) + \frac{\partial g}{\partial y}(0,0)(y-0) \\ &= 0 + 0(x-0) + 0(y-0) = 0 \end{aligned}$$

i.e., $L_{(0,0)}(x,y) = 0$ (so it's the xy-plane). However, $L_{(0,0)}$ is not, in any sense, an approximation of g. For instance, at all points on the line $y = x$, except at $(0,0)$,

$$g(x,y) = \frac{x^2}{x^2+x^2} = \frac{1}{2}$$

So no matter how close to $(0,0)$ we get (i.e., no matter how small an open disk around $(0,0)$ we take), g will be equal to $1/2$ at some point inside that disk.

If $(x,y) \neq (0,0)$, then

$$\frac{\partial g}{\partial x}(x,y) = \frac{y(x^2+y^2) - xy\cdot 2x}{(x^2+y^2)^2} = \frac{y^3 - x^2y}{(x^2+y^2)^2}$$

Earlier, we found that $\partial g/\partial x(0,0) = 0$. We now show that $\partial g/\partial x$ is not continuous at $(0,0)$; i.e., we show that

$$\lim_{(x,y)\to(0,0)} \frac{\partial g}{\partial x}(x,y) \neq \frac{\partial g}{\partial x}(0,0) \tag{5.1}$$

Consider the path toward $(0,0)$ along $y = 2x$:

$$\begin{aligned} \lim_{(x,y)\to(0,0)} \frac{\partial g}{\partial x}(x,y) &= \lim_{(x,y)\to(0,0)} \frac{y^3 - x^2y}{(x^2+y^2)^2} \\ &= \lim_{(x,y)\to(0,0)} \frac{(2x)^3 - x^2(2x)}{(x^2+(2x)^2)^2} \\ &= \lim_{(x,y)\to(0,0)} \frac{6x^3}{25x^4} \\ &= \frac{6}{25}\lim_{(x,y)\to(0,0)} \frac{1}{x} \end{aligned}$$

which is not a real number. So, indeed, $\partial g/\partial x$ is not continuous at $(0,0)$. In the same way, we show that $\partial g/\partial y$ is not continuous at $(0,0)$. ▲

To summarize the previous example:

(1) With the function $g(x,y)$, we *can* associate a plane (by using the linearization formula from Definition 12), but that plane is not an approximation of g near $(0,0)$.

(2) The partial derivatives $\partial g/\partial x$ and $\partial g/\partial y$ are not continuous at $(0,0)$.

What can we conclude from this? For now—only that by requiring that partial derivatives be continuous we *might* be able to eliminate the situation that we saw in Example 5.3. In other words, continuity of partial derivatives might guarantee that we will obtain a meaningful linear approximation. (As a matter of fact this is true, but we need some work to get there.)

Differentiability and Linear Approximation

Let us go back to the one-variable case. Assuming that $y = f(x)$ is differentiable at a, its linearization is given by $L_a(x) = f(a) + f'(a)(x - a)$.

In Example 5.1 we computed that $L_1(x) = 2x + 1$ is the linearization of $f(x) = x + 2\sqrt{x}$ at $a = 1$. So

$$f(x) \approx L_1(x) = 2x + 1$$

for x near 1. For instance, $f(0.99) = 2.97997$ is approximated by $L_1(0.99) = 2.98$ and $f(1.04) = 3.07961$, whereas $L_1(1.04) = 3.08$.

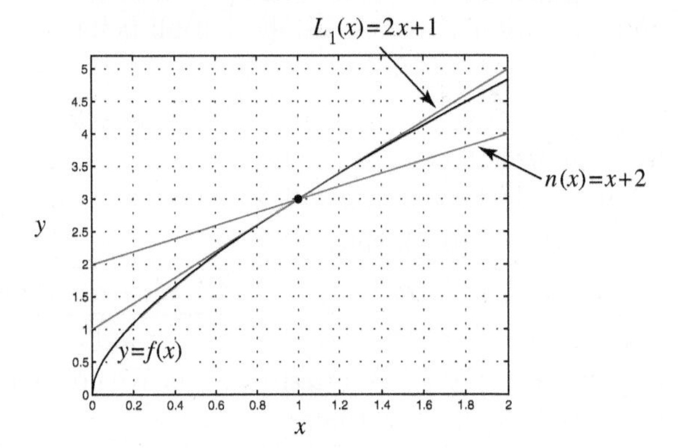

FIGURE 5.5
Lines approximating the function

Now consider the line $n(x) = x+2$; see Figure 5.5. Clearly, $n(1) = 3$. The values $n(0.99) = 2.99$ and $n(1.04) = 3.04$ approximate the values $f(0.99) = 2.97997$ and $f(1.04) = 3.07961$ quite well. If we get closer, we get even better approximation: for instance, $n(0.9999) = 2.9999$ closely approximates the value $f(0.9999) = 2.99980$.

So the fact that for x near $a = 1$, the function $f(x)$ and its linearization $L_a(x)$ have approximately the same values is nothing special. As a matter of fact, any line (except the vertical line) that goes through the point of tangency $(1, 3)$ shares the same property. Clearly, we need to investigate further to find what makes $L_a(x)$, i.e., the tangent line, special.

The quantity $|f(x) - L_a(x)|$ measures the absolute value of the error; it says by how much the linear approximation L_a of f differs from f. We investigate the fraction

$$\frac{|f(x) - L_a(x)|}{|x - a|}$$

which compares the error to the distance (how far x is from a); see Figure 5.6.

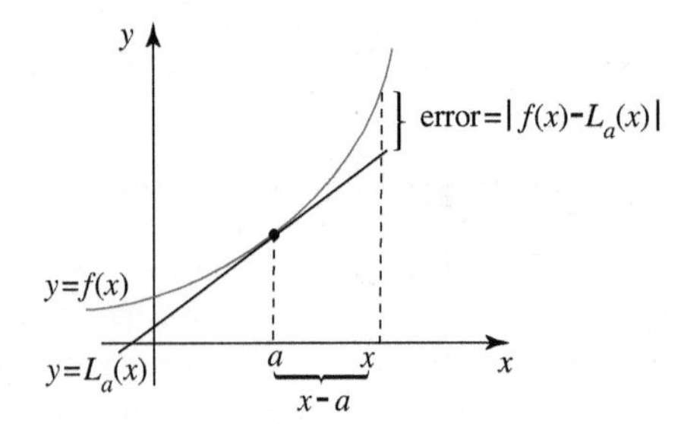

FIGURE 5.6
Analyzing linearization

Assuming that $x \neq a$,

$$\frac{|f(x) - L_a(x)|}{|x-a|} = \left|\frac{f(x) - (f(a) + f'(a)(x-a))}{x-a}\right|$$
$$= \left|\frac{f(x) - f(a)}{x-a} - \frac{f'(a)(x-a)}{x-a}\right|$$
$$= \left|\frac{f(x) - f(a)}{x-a} - f'(a)\right| \qquad (5.2)$$

Note that (take $h = x - a$; then, as $x \to a$, $h \to 0$)

$$\lim_{x \to a} \frac{f(x) - f(a)}{x - a} = \lim_{h \to 0} \frac{f(a+h) - f(a)}{h} = f'(a)$$

Taking the limit in (5.2) we get

$$\lim_{x \to a} \frac{|f(x) - L_a(x)|}{|x-a|} = |f'(a) - f'(a)| = 0 \qquad (5.3)$$

What does (5.3) say?

We know that as $x \to a$, $|f(x) - L_a(x)| \to 0$ and $|x - a| \to 0$. Formula (5.3) says that the error

$$|f(x) - L_a(x)|$$

approaches 0 *more quickly* than the distance $|x - a|$ does.

Example 5.4 Comparing Errors

In Example 5.1 we showed that $L_1(x) = 2x + 1$ is the linearization of the function $f(x) = x + 2\sqrt{x}$ at $a = 1$. Using L'Hôpital's rule,

$$\lim_{x \to 1} \frac{|f(x) - L_1(x)|}{|x-1|} = \lim_{x \to 1} \left|\frac{(x + 2\sqrt{x}) - (2x+1)}{x-1}\right|$$
$$= \lim_{x \to 1} \left|\frac{-x + 2\sqrt{x} - 1}{x-1}\right|$$
$$= \lim_{x \to 1} \left|\frac{-1 + 1/\sqrt{x}}{1}\right| = 0$$

So the error $|f(x) - L_1(x)|$ approaches zero more quickly than $|x - 1|$ does.

Now we do the same for $n(x) = x + 2$; again, using L'Hôpital's rule,

$$\lim_{x \to 1} \frac{|f(x) - n(x)|}{|x-1|} = \lim_{x \to 1} \left|\frac{(x + 2\sqrt{x}) - (x+2)}{x-1}\right|$$
$$= \lim_{x \to 1} \left|\frac{2\sqrt{x} - 2}{x-1}\right|$$
$$= \lim_{x \to 1} \left|\frac{1/\sqrt{x}}{1}\right| = 1$$

In this case, the error $|f(x) - n(x)|$ approaches zero at the same rate as $|x - 1|$ does.

What we have just shown could be proven in general: of all lines going through $(a, f(a))$, the linearization has the error that approaches zero most quickly. And that is what distinguishes linearization from any other line.

The formula (5.3) is what we can use to generalize the notion of differentiability to functions of two variables. Note that the expression in (5.3) is the absolute value of the error divided by the distance.

Definition 13 Differentiability for a Function of Two Variables

Assume that f is a real-valued function of two variables and let

$$L_{(a,b)}(x,y) = f(a,b) + f_x(a,b)(x-a) + f_y(a,b)(y-b)$$

We say that f is *differentiable* at (a,b) if

(a) both partial derivatives f_x and f_y exist in some disk around (a,b).

(b) the function $L_{(a,b)}(x,y)$ satisfies

$$\lim_{(x,y)\to(a,b)} \frac{|f(x,y) - L_{(a,b)}(x,y)|}{\sqrt{(x-a)^2 + (y-b)^2}} = 0$$

So for a differentiable function f, the linear function $z = L_{(a,b)}(x,y)$ is the plane tangent to the graph of f at (a,b). The linear approximation is a good approximation for f near (a,b) in the sense given in (b). In other words, if f is differentiable, then

$$f(x,y) \approx L_{(a,b)}(x,y) = f(a,b) + f_x(a,b)(x-a) + f_y(a,b)(y-b)$$

for (x,y) near (a,b).

The condition (b) in Definition 13 is not easy to verify. However, there is an easier set of conditions that guarantee differentiability of a function. Earlier, in Example 5.3, we saw that the mere existence of partial derivatives does not suffice. The following theorem (which we present without proof) tells us that if the partial derivatives are continuous, then differentiability is guaranteed.

Theorem 6 Sufficient Condition for Differentiability

Assume that f is defined on an open disk $B_r(a,b)$ centred at (a,b) and that the partial derivatives f_x and f_y are continuous on $B_r(a,b)$. Then f is differentiable at (a,b).

Note that the phrase "on an open disk $B_r(a,b)$ centred at (a,b)" is a precise translation of the phrase "near (a,b)."

The word "sufficient" in the title of Theorem 6 means that the conditions in the theorem guarantee that f is differentiable, but f could be differentiable under weaker conditions as well. (A weaker condition is one that is satisfied by more objects than the original condition; for instance: the fact that a number ends with 5 is a sufficient condition for it to be divisible by 5. A weaker condition—a number ends with 5 or with 0 — guarantees divisibility by 5 as well.)

As in the case of one variable, the following is true.

Theorem 7 Differentiability Implies Continuity

Assume that a function f is differentiable at (a,b). Then it is continuous at (a,b) as well.

Example 5.5 Linearization and Linear Approximation

Show that

$$\frac{xy}{x+3y-2} \approx \frac{3}{8} + \frac{21}{64}(x-1) - \frac{1}{64}(y-3)$$

for (x,y) near $(1,3)$.

▶ Let

$$f(x,y) = \frac{xy}{x+3y-2}$$

5. The contour diagram of the function $f(x,y) = \sqrt{x^2 + y^2}$ consists of concentric circles; see Example 2.8 in Section 2. Describe what happens as you zoom in on the level curves near $(0,0)$. What does that mean for the function f?

6. What is the differential of the function $f(x,y) = 4$? Does it make sense?

7–10 ▪ Verify that the following approximations are valid for (x,y) near the specified point.

7. $\ln(x^2 - y^2) \approx 2x - 2$; $(1,0)$

8. $\dfrac{1}{xy} \approx \dfrac{3}{2} - \dfrac{1}{2}x - \dfrac{1}{4}y$; $(1,2)$

9. $\sqrt{x^2 + 4y^2} \approx \dfrac{3}{5}x + \dfrac{8}{5}y$; $(3,2)$

10. $\arctan(y/x) \approx y$; $(1,0)$

11. Using Theorem 6, show that the function $f(x,y) = x^2ye^y - 2$ is differentiable at all (x,y) in $\mathbb{R}^2$.

12. Using Theorem 6, show that the function $f(x,y) = e^{y-x}$ is differentiable at all (x,y) in $\mathbb{R}^2$.

13. Explain why the function $f(x,y) = x^2y - y^3$ is differentiable at $(1,5)$. Find the linearization of f at $(1,5)$.

14. Explain why the function $f(x,y) = xy(x^2 + y^2)^{-1}$ is differentiable at $(1,-1)$. Find the linearization of f at $(1,-1)$.

15. Using Theorem 6, show that the function $f(x,y) = \ln(x^2 + y^2)$ is differentiable at $(1,1)$. What is the largest open disk (refer to the statement of the theorem) centred at $(1,1)$ that you can use?

16. Using Theorem 6, show that the function $f(x,y) = x \tan y$ is differentiable at $(0,0)$. What is the largest open disk (refer to the statement of the theorem) centred at $(0,0)$ that you can use?

17–22 ▪ Find the equation of the plane tangent to the given surface at the point indicated.

17. $f(x,y) = 2x^2 - xy + 4y^3 - 3$; $(1,0,-1)$

18. $g(x,y) = \dfrac{xy}{x - 2y}$; $(3,1,3)$

19. $g(x,y) = 3ye^{x+y}$; $(-2,2,6)$

20. $f(x,y) = \sqrt{xy + 7}$; $(6,7,7)$

21. $f(x,y) = x + \ln(x^2 + y + 4)$; $(0,0,\ln 4)$

22. $f(x,y) = \arctan(2y/x)$; $(2,1,\pi/4)$

23. Find the linear approximation of the function $f(x,y) = ye^{-x^2}$ at $(0,1)$ and use it to find an approximation of $f(-0.1, 0.9)$.

24–27 ▪ Approximate the value of the given expression and compare with the calculator value.

24. $8.04^2 \ln 1.05$

25. $\sqrt{2.98^2 + 4.04^2}$

26. $\sin 1.5 \cos 0.1$

27. $2.98e^{-0.04}$

28. Consider the body surface area $S(m,h) = 0.20247m^{0.425}h^{0.725}$ of a human of height h (metres) and mass m (kilograms). Find the differential dS when $m = 65$ kg, $h = 1.55$ m, $dm = 0.5$ kg, and $dh = 0.03$ m, and interpret your answer.

29. Consider the body surface area $S(m,h) = 0.20247m^{0.425}h^{0.725}$ of a human of height h (metres) and mass m (kilograms). If the mass increases by 3% and the height by 2%, by how much does the surface area increase? Find a general formula, and then calculate in the case $m = 75$ kg and $h = 1.72$ m.

30. Recall that $V(a,b) = ab(a+b)\pi/12$ gives an estimate for the volume of a tumour based on the elliptic image showing in a mammogram (a and b are perpendicular diameters); see Example 1.7 in Section 1. Find the differential dV when $a = 8$, $b = 3.4$, $da = 0.4$, and $db = 0.6$ (all in millimetres), and interpret your answer.

31. Consider the formula $V(a,b) = ab(a+b)\pi/12$ which gives an estimate for the volume of a tumour (a and b are perpendicular diameters measured in a mammogram); see Example 1.7 in Section 1. In the first mammogram, the measurements are $a = 6$ mm and $b = 4.4$ mm. A second mammogram shows an

increase of 0.2 mm in the direction of the larger diameter and 0.3 mm in the direction of the smaller diameter. Using differentials, estimate the increase in the volume of the tumour.

32. The pressure in kilopascals (kPa) in an ideal gas is given by $P(T,V) = RnT/V$, where V is the volume (in litres) occupied by the gas, T is the absolute temperature (in degrees Kelvin), n is the number of moles of gas, and $R = 8.314 \text{ J mol}^{-1}\text{ K}^{-1}$ is the universal gas constant. Compute the differential of P and explain the signs of the coefficients of dT and dV.

33. About how accurately can the volume of a cylinder be calculated from measurements of its height and radius that are in error by 2.5%?

34. Let $f(x,y) = 4x^3y^2$. Estimate the change in the function f if x increases by 4% and y decreases by 3%.

35. Estimate the change in the function $f(x,y) = cx^2y^3$ (c is a non-zero constant) if x decreases by 2% and y increases by 2%.

36. Consider the function

$$f(x,y) = \begin{cases} 1 & \text{if } xy = 0 \\ 0 & \text{if } xy \neq 0 \end{cases}$$

Assuming that the partial derivatives of f are continuous at $(0,0)$, find the linearization of f at $(0,0)$. Explain why the linearization you obtained does not make sense in terms of approximating f near $(0,0)$. Are the partial derivatives of f continuous at $(0,0)$?

37. Consider the function

$$f(x,y) = \begin{cases} \dfrac{xy}{x^2+y^2} & \text{if } (x,y) \neq (0,0) \\ 0 & \text{if } (x,y) = (0,0) \end{cases}$$

Ignoring the assumption on the continuity of partial derivatives, find the linearization of f at $(0,0)$ anyway. Explain why the linearization you obtained does not make sense in terms of approximating f near $(0,0)$. What can you conclude about the continuity of the partial derivatives of f at $(0,0)$?

6 The Chain Rule

In this section, we learn how to differentiate **compositions of functions.** We introduce **curves and surfaces defined implicitly,** and use the chain rule to calculate the derivatives of an **implicitly defined function,** i.e., a function defined by an equation.

Chain Rule

To calculate the derivative of a composition of functions, we need the chain rule. For instance, if $z = f(x, y)$ is a function of x and y, and both x and y depend on t (using symbols, we write $x = x(t)$ and $y = y(t)$), then z is a function of t as well:

$$z = f(x(t), y(t))$$

Or, it could happen that x and y depend on two variables, $x = x(u, v)$ and $y = y(u, v)$. In that case, z is a function of u and v:

$$z = f(x(u, v), y(u, v))$$

How do we calculate the derivatives of z in these, and similar, cases?

To start, consider the following example.

Example 6.1 Dynamics of Change in Wolf Population

The number of wolves W in a certain fixed region (habitat) depends on both the availability of food F and the distance L from urban areas.

Assume that $\partial W/\partial F = 0.4$ (i.e., the wolf population is an increasing function of the availability of food) and $\partial W/\partial L = 0.6$ (this means that the number of wolves increases with increasing distance from urban areas; in other words, W decreases as the distance between urban areas and the wolves' habitat decreases). Over time, the availability of food decreases at a rate of 2 units per year, and the urban areas grow, shortening the distance to the wolves' habitat by 0.4 km per year.

We would like to estimate the current change dW/dt in the population of wolves. To summarize: the wolf population W depends on the food F and on the distance from the human population L, both of which change with time. Thus, W depends on the time t, and we are asked to find the rate of change of W with respect to t.

Let ΔW denote the change in the wolf population over the time interval Δt. Using formula (5.6) from Section 5, we write

$$\Delta W \approx dW = \frac{\partial W}{\partial F}\Delta F + \frac{\partial W}{\partial L}\Delta L = 0.4\Delta F + 0.6\Delta L$$

It is given that $F'(t) = -2$, and since (see formula (5.4) in Section 5)

$$\Delta F \approx dF = F'(t)dt$$

we get $\Delta F \approx -2dt = -2\Delta t$. As well, from

$$\Delta L \approx dL = L'(t)dt$$

we get $\Delta L \approx -0.4dt = -0.4\Delta t$. Thus,

$$\Delta W \approx 0.4(-2\Delta t) + 0.6(-0.4\Delta t) = -1.04\Delta t$$

and

$$\frac{\Delta W}{\Delta t} \approx -1.04.$$

Taking the limit as $\Delta t \to 0$, we get that $dW/dt \approx -1.04$. Thus, the wolf population decreases at a rate of about one wolf per year. ▲

Guided by the reasoning in the previous example, we now state the chain rule.

Theorem 8 Chain Rule

Assume that $z = f(x, y)$ is a differentiable function of two variables x and y and $x = x(t)$ and $y = y(t)$ are differentiable functions of t. Then $z = f(x(t), y(t))$ is a differentiable function of t, and

$$z'(t) = \frac{\partial f}{\partial x}x'(t) + \frac{\partial f}{\partial y}y'(t).$$

Using differential notation for the derivatives, we write

$$\frac{dz}{dt} = \frac{\partial f}{\partial x}\frac{dx}{dt} + \frac{\partial f}{\partial y}\frac{dy}{dt}.$$

Notational convention: we use d/d for the derivatives of functions of one variable, and ∂/∂ for the partial derivatives of functions of several variables.

Sometimes, we replace f by z in the chain rule formula and write

$$\frac{dz}{dt} = \frac{\partial z}{\partial x}\frac{dx}{dt} + \frac{\partial z}{\partial y}\frac{dy}{dt} \tag{6.1}$$

In formula (6.1) we use the same letter, z, for two different functions: z as a function of t (left side) and z as a function of the variables x and y (right side). This common practice, now that we are aware ot it, will not be a cause of confusion.

For instance, let $z = 3x^2e^y$, $x = \cos t$, and $y = 5t$. The function z on the right side in (6.1) is $z(x, y) = 3x^2e^y$; on the left side, $z(t) = 3x^2e^y = 3\cos^2 t\, e^{5t}$.

We now sketch the proof of the chain rule. As in Example 6.1, we estimate Δz and then use the fact that

$$\frac{dz}{dt} = \lim_{\Delta t \to 0} \frac{\Delta z}{\Delta t}$$

Viewing z as a function of x and y, we write (see (5.6) in Section 5)

$$\Delta z \approx \frac{\partial f}{\partial x}\Delta x + \frac{\partial f}{\partial y}\Delta y$$

Using the linear approximations $\Delta x \approx dx = x'(t)\Delta t$ and $\Delta y \approx dy = y'(t)\Delta t$, we obtain

$$\Delta z \approx \frac{\partial f}{\partial x}x'(t)\Delta t + \frac{\partial f}{\partial y}y'(t)\Delta t$$

and

$$\frac{\Delta z}{\Delta t} \approx \frac{\partial f}{\partial x}x'(t) + \frac{\partial f}{\partial y}y'(t)$$

As $\Delta t \to 0$,

$$\frac{dz}{dt} = \lim_{\Delta t \to 0} \frac{\Delta z}{\Delta t} = \frac{\partial f}{\partial x}x'(t) + \frac{\partial f}{\partial y}y'(t)$$

and we are done.

Example 6.2 Chain Rule

Let $z = 3x^2e^y$, $x = \cos t$, and $y = 5t$. Find $z'(t)$ directly, and then using the chain rule.

▶ To calculate $z'(t)$ directly, we find the composition first

$$z(t) = 3x^2e^y = 3\cos^2 t\, e^{5t}$$

and then differentiate

$$z'(t) = 3(2\cos t)(-\sin t)\, e^{5t} + 3\cos^2 t\, e^{5t}(5) = -6\cos t \sin t\, e^{5t} + 15\cos^2 t\, e^{5t}$$

Alternatively, we use the chain rule (Theorem 8)

$$\begin{aligned}\frac{dz}{dt} &= \frac{\partial f}{\partial x}\frac{dx}{dt} + \frac{\partial f}{\partial y}\frac{dy}{dt}\\ &= (6xe^y)(-\sin t) + (3x^2e^y)(5)\\ &= (6\cos t\, e^{5t})(-\sin t) + 3(\cos t)^2\, e^{5t}(5)\\ &= -6\cos t \sin t\, e^{5t} + 15\cos^2 t\, e^{5t}\end{aligned}$$

An easy way to remember the chain rule formula (and to generalize it, if needed) is to draw a tree diagram showing how the variables depend on each other.

Suppose that $z = f(x, y)$, $x = x(t)$, and $y = y(t)$ and we wish to find $z'(t)$. We place the function whose derivative we are calculating on the top, and draw a branch leading to each variable; thus, we place z on top and draw branches from $f(x, y)$ to its variables x and y, and then one branch from each of x and y to t; see Figure 6.1a. Along each branch, we write the corresponding derivative (using ∂/∂ for partial derivatives if a function has several variables, and d/d for derivatives of functions of one variable).

According to the chain rule, we multiply the derivatives along each path from f to t and add up the products:

$$\frac{dz}{dt} = \frac{\partial f}{\partial x}\frac{dx}{dt} + \frac{\partial f}{\partial y}\frac{dy}{dt}.$$

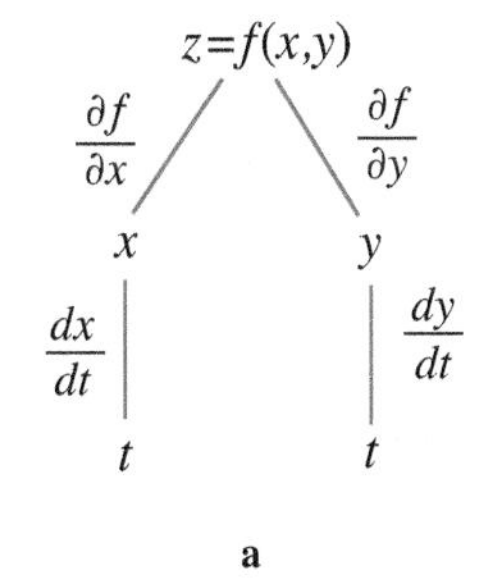

a

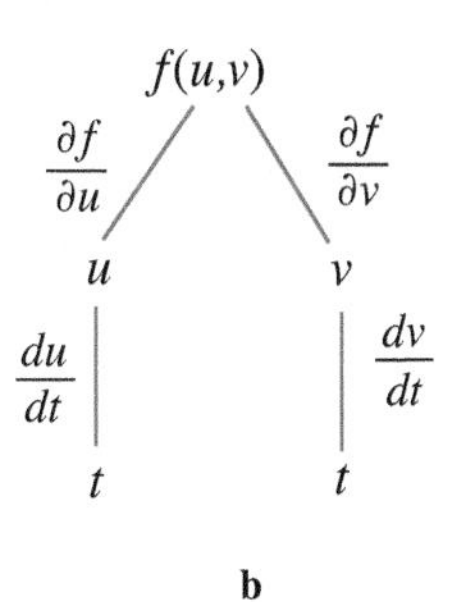

b

FIGURE 6.1

Tree diagram for the chain rule

Example 6.3 Chain Rule

Let $f(u, v) = v^3 \ln u$, where $u = 4t^2$ and $v = t + 4$. Find $f'(t)$.

▶ Using the tree diagram in Figure 6.1b, we write

$$\begin{aligned}f'(t) = \frac{df}{dt} &= \frac{\partial f}{\partial u}\frac{du}{dt} + \frac{\partial f}{\partial v}\frac{dv}{dt}\\ &= \frac{v^3}{u}\cdot 8t + 3v^2 \ln u \cdot 1\\ &= \frac{(t+4)^3}{4t^2}\cdot 8t + 3(t+4)^2 \ln(4t^2)\\ &= \frac{2(t+4)^3}{t} + 3(t+4)^2\ln(4t^2)\end{aligned}$$

Example 6.4 Interpreting the Chain Rule Formula

The production S of soybeans in the prairie provinces in Canada depends on the average yearly temperature T and the total annual rainfall R; thus, $S = f(T, R)$. Over the last decade, it has been determined that the average temperature increases at a rate of about 0.1°C per year, and the average rainfall decreases by about 0.08 cm per year.

Both T and R change with time, t, and thus S is a function of time as well, i.e.,

$$S(t) = f(T(t), R(t))$$

By the chain rule,

$$\frac{dS}{dt} = \frac{\partial S}{\partial T}\frac{dT}{dt} + \frac{\partial S}{\partial R}\frac{dR}{dt}$$

Explain the meaning of each of the derivatives on the right side and comment on their sign.

▶ The partial derivative $\partial S/\partial T$ is the rate of change in the soybean production with respect to the average temperature (assuming that the average rainfall remains constant). No information about how it changes is given. If the production increases with increased average temperature, then $\partial S/\partial T > 0$; otherwise, if an increase in the average temperature causes a decline in the production, then $\partial S/\partial T < 0$.

The derivative dT/dt is the rate of change of average yearly temparature over time and is given: $dT/dt = 0.1 > 0$.

The partial derivative $\partial S/\partial R$ describes how the soybean production depends on the rainfall (again, there is no information about the change in the text of the problem). If increased rainfall (with no change in the average temperature) increases the production of soybeans, then $\partial S/\partial R > 0$. However, if the production shows a decrease when there is an increase in average rainfall (and the average temperature does not change), then $\partial S/\partial R < 0$.

Finally, dR/dt is the time rate of change of average rainfall. It is assumed that it decreases over the years, at a rate of $dR/dt = -0.08 < 0$. ▲

Now assume that $z = f(x, y)$ is a differentiable function of x and y, and each of x and y is a differentiable function of u and v. Thus,

$$z = f(x(u, v), y(u, v))$$

i.e., z is a function of u and v. How do we calculate its partial derivatives $\partial z/\partial u$ and $\partial z/\partial v$?

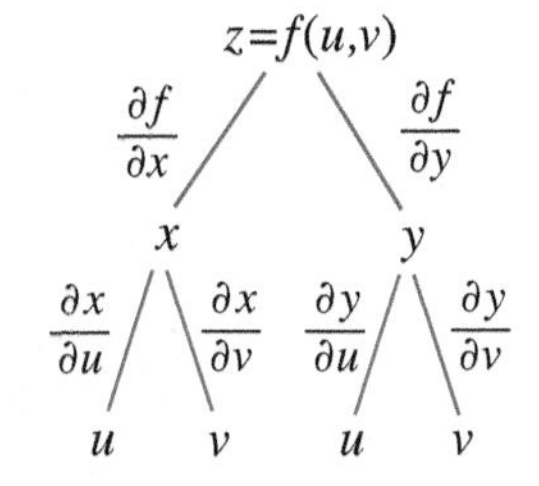

FIGURE 6.2

Tree diagram for the composition $z = f(x(u, v), y(u, v))$

To find $\partial z/\partial u$ we hold v fixed and view z as a function of *one* variable u. We apply the chain rule (Theorem 8) and obtain

$$\frac{\partial z}{\partial u} = \frac{\partial f}{\partial x}\frac{\partial x}{\partial u} + \frac{\partial f}{\partial y}\frac{\partial y}{\partial u} \tag{6.2}$$

(note that we replaced all d/d with ∂/∂ since all functions involved depend on two variables). This is exactly what we obtain from the tree diagram; see Figure 6.2. Looking at the diagram, we obtain

$$\frac{\partial z}{\partial v} = \frac{\partial f}{\partial x}\frac{\partial x}{\partial v} + \frac{\partial f}{\partial y}\frac{\partial y}{\partial v}$$

Example 6.5 **Chain Rule**

Let $f(x, y) = x^2y^3$, $x = 2u + v^2$, and $y = u^2 - v$. Find $\partial f/\partial u$.

▶ By the chain rule (6.2),

$$\begin{aligned}\frac{\partial f}{\partial u} &= \frac{\partial f}{\partial x}\frac{\partial x}{\partial u} + \frac{\partial f}{\partial y}\frac{\partial y}{\partial u}\\ &= 2xy^3(2) + (3x^2y^2)(2u)\\ &= 4(2u + v^2)(u^2 - v)^3 + 6u(2u + v^2)^2(u^2 - v)^2\\ &= 2(2u + v^2)(u^2 - v)^2\left[2(u^2 - v) + 3u(2u + v^2)\right]\end{aligned}$$

▲

Implicit Differentiation

Recall that we have already used implicit differentiation in single-variable calculus, where the function y was given implicitly as a function of x, such as in

$$x^2 + y^2 = 4$$

To compute dy/dx, we differentiated this equation with respect to x, keeping in mind that y is a function of x:

$$2x + 2y\frac{dy}{dx} = 0$$
$$\frac{dy}{dx} = -\frac{x}{y}$$

We can write the equation $x^2 + y^2 = 4$ in the form

$$F(x, y) = 0$$

where $F(x, y) = x^2 + y^2 - 4$ is a function of *two* variables.

Definition 14 Implicitly Defined Curve

Assume that $F(x, y)$ is a differentiable function such that its partial derivatives $\partial F/\partial x$ and $\partial F/\partial y$ are continuous. The set of points (x, y) in the domain of F for which $F(x, y) = 0$ is called an *implicitly defined curve* (or a *curve defined implicitly*).

Recall that a level curve of a differentiable function $f(x, y)$ of value c (Section 2) is given by the equation

$$f(x, y) = c$$

i.e., it is a curve defined implicitly. To see that it fits into Definition 14, we write $f(x, y) - c = 0$ and take $F(x, y) = f(x, y) - c$.

The curve defined implicitly by $F(x, y) = x^2 + y^2 - 4 = 0$ is a circle of radius 2 centred at the origin (note that both partial derivatives $\partial F/\partial x = 2x$ and $\partial F/\partial y = 2y$ are continuous).

Curves defined implicitly possess a wide range of properties. For instance, the curve defined by

$$F_1(x, y) = (x^2 + y^2)^2 - x^2 + y^2 = 0$$

has a self-intersection; see Figure 6.3a. It is called a *lemniscate* (the infinity sign, also known as the *figure-eight curve*).

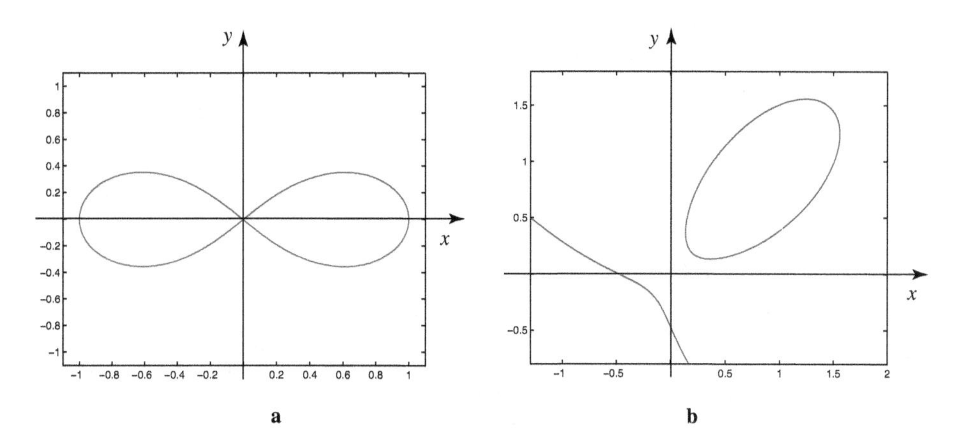

FIGURE 6.3
Curves defined implicitly

The curve defined by

$$F_2(x, y) = x^3 + y^3 - 3xy + 0.1 = 0$$

consists of two separate pieces; see Figure 6.3b. In Exercises 7 to 10 we examine additional examples.

In some cases, the equation $F(x, y) = 0$ can be solved to express y explicitly as a function of x. For instance, solving $x^2 + y^2 - 4 = 0$ we obtain two curves

$$y = \sqrt{4 - x^2}$$

(upper semi-circle) and

$$y = -\sqrt{4 - x^2}$$

(lower semi-circle).

In general, solving for y explicitly is difficult (such as solving $(x^2 + y^2)^2 - x^2 + y^2 = 0$ or $x^3 + y^3 - 3xy + 0.1 = 0$ for y) or impossible (for instance solving $xy + e^y = 0$ for y). Nevertheless, in *all cases* we can find the derivative dy/dx. We now show how.

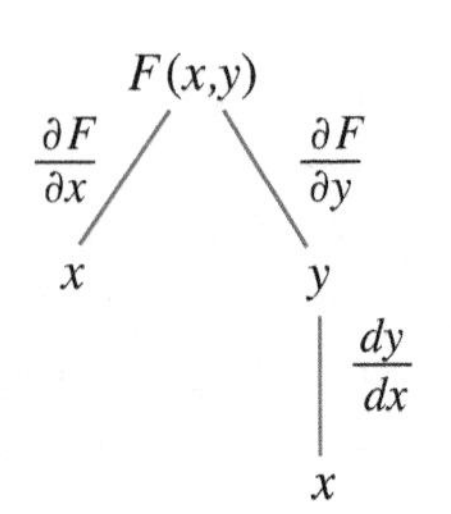

FIGURE 6.4

Tree diagram for the chain rule

Consider $F(x, y) = 0$, and assume that x is an independent variable and y is a function of x. Then F is a function of x (see the diagram in Figure 6.4) and, by the chain rule,

$$\frac{dF}{dx} = \frac{\partial F}{\partial x} + \frac{\partial F}{\partial y}\frac{dy}{dx} = 0$$

Thus, assuming that $\partial F/\partial y \neq 0$,

$$\frac{dy}{dx} = -\frac{\partial F/\partial x}{\partial F/\partial y}$$

(So, deep down, the implicit differentiation of a function of one variable is the chain rule applied to a function of two variables.)

In the case of the circle, $F(x, y) = x^2 + y^2 - 4$, $\partial F/\partial x = 2x$, and $\partial F/\partial y = 2y$, and thus

$$\frac{dy}{dx} = -\frac{\partial F/\partial x}{\partial F/\partial y} = -\frac{2x}{2y}$$

as long as $y \neq 0$. (Note that when $y = 0$, $x = \pm 2$; thus, the denominator is zero at the points where the circle crosses the x-axis, and the tangent to the circle is vertical.)

Theorem 9 Implicit Derivative

Assume that $F(x, y) = 0$ defines y implicitly as a function of x, where F is differentiable and its partial derivatives $\partial F/\partial x$ and $\partial F/\partial y$ are continuous. Then

$$\frac{dy}{dx} = -\frac{\partial F/\partial x}{\partial F/\partial y}$$

provided that $\partial F/\partial y \neq 0$.

Moving one dimension higher, we study implicitly defined surfaces.

Definition 15 Implicitly Defined Surface

Assume that $F(x, y, z)$ is a differentiable function such that its partial derivatives $\partial F/\partial x$, $\partial F/\partial y$, and $\partial F/\partial z$ are continuous. The set of points (x, y, z) in the domain of F for which $F(x, y, z) = 0$ is called an *implicitly defined surface* (or a *surface defined implicitly*).

For instance, the equation

$$F(x, y, z) = x^2 + y^2 + z^2 - 4 = 0$$

defines a sphere of radius 2 centred at the origin (note that all partial derivatives of F are continuous). This equation explicitly defines two surfaces: the upper hemisphere

$$z = \sqrt{4 - x^2 - y^2}$$

and the lower hemisphere

$$z = -\sqrt{4 - x^2 - y^2}$$

Differentiating $F(x, y, z) = x^2 + y^2 + z^2 - 4 = 0$ implicitly, we can calculate the partial derivatives of z. For $\partial z/\partial x$, we view y as constant:

$$2x + 0 + 2z\frac{\partial z}{\partial x} = 0$$

$$\frac{\partial z}{\partial x} = -\frac{x}{z}$$

Likewise, keeping x fixed and viewing z as a function of y,

$$0 + 2y + 2z\frac{\partial z}{\partial y} = 0$$

$$\frac{\partial z}{\partial y} = -\frac{y}{z}$$

Now we do it in general: assume that the equation $F(x, y, z) = 0$ implicitly defines the surface $z = f(x, y)$. Using the chain rule, holding y constant, and with the help of the tree diagram in Figure 6.5, we get

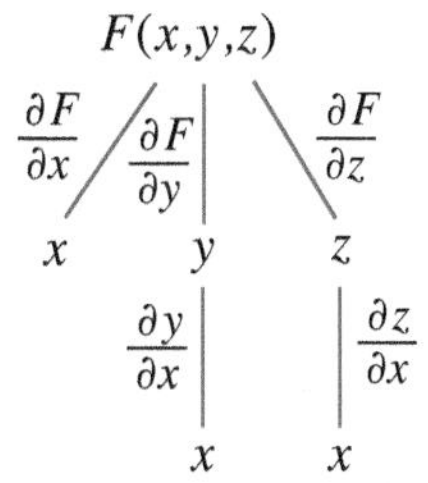

$$\frac{\partial F}{\partial x} + \frac{\partial F}{\partial y}\frac{\partial y}{\partial x} + \frac{\partial F}{\partial z}\frac{\partial z}{\partial x} = 0$$

$$\frac{\partial F}{\partial x} + \frac{\partial F}{\partial y}(0) + \frac{\partial F}{\partial z}\frac{\partial z}{\partial x} = 0$$

and

$$\frac{\partial z}{\partial x} = -\frac{\partial F/\partial x}{\partial F/\partial z}$$

provided that $\partial F/\partial z \neq 0$. We calculate $\partial z/\partial y$ in exactly the same way.

FIGURE 6.5

Tree diagram for implicit differentiation

Theorem 10 Implicit Derivative

Assume that $F(x, y, z) = 0$ defines z implicitly as a function of x and y, where F is differentiable and its partial derivatives $\partial F/\partial x$, $\partial F/\partial y$, and $\partial F/\partial z$ are continuous. Then

$$\frac{\partial z}{\partial x} = -\frac{\partial F/\partial x}{\partial F/\partial z} \quad \text{and} \quad \frac{\partial z}{\partial y} = -\frac{\partial F/\partial y}{\partial F/\partial z}$$

assuming that $\partial F/\partial z \neq 0$.

Example 6.6 Implicit Differentiation

Find $\partial z/\partial x$ and $\partial z/\partial y$ if $x^2y + e^{yz} = 1$.

Define $F(x, y, z) = x^2y + e^{yz} - 1$. We have a choice: we can proceed by implicitly differentiating the equation $F(x, y, z) = 0$ or use the formulas provided by Theorem 10. For practice, we do it both ways.

Thinking of z as a function of x, and differentiating $x^2y + e^{yz} - 1 = 0$, we get

$$2xy + ye^{yz}\frac{\partial z}{\partial x} = 0$$

$$\frac{\partial z}{\partial x} = -\frac{2xy}{ye^{yz}} = -\frac{2x}{e^{yz}}$$

Likewise, viewing z as a function of y, and keeping x fixed, we get

$$x^2 + e^{yz}\left(z + y\frac{\partial z}{\partial y}\right) = 0$$

$$ye^{yz}\frac{\partial z}{\partial y} = -x^2 - ze^{yz}$$

$$\frac{\partial z}{\partial y} = -\frac{x^2 + ze^{yz}}{ye^{yz}}$$

Alternatively, thinking of F as a function of three variables, we have

$$\frac{\partial F}{\partial x} = 2xy$$

$$\frac{\partial F}{\partial y} = x^2 + ze^{yz}$$

$$\frac{\partial F}{\partial z} = ye^{yz}$$

By Theorem 10,

$$\frac{\partial z}{\partial x} = -\frac{\partial F/\partial x}{\partial F/\partial z} = -\frac{2xy}{ye^{yz}} = -\frac{2x}{e^{yz}}$$

and

$$\frac{\partial z}{\partial y} = -\frac{\partial F/\partial y}{\partial F/\partial z} = \frac{x^2 + ze^{yz}}{ye^{yz}}$$

▲

Example 6.7 Implicit Differentiation and the Chain Rule

The ideal gas law states that

$$PV = nRT$$

where P is the pressure (in pascals), V is the volume (in cubic metres), and T is the absolute temperature (in degrees Kelvin). The symbol n denotes the quantity of the substance (in moles) and R is the universal gas constant ($R = 8.314$ in SI units). We can rewrite the above equation as $F(P, V, T) = 0$, where

$$F(P, V, T) = PV - nRT$$

If we wish to study how the volume changes, then we view $F(P, V, T) = 0$ as defining V implicitly in terms of P and T (in this case, of course, we can solve explicitly for V; however, our aim here is to show how implicit differentiation works). If we wish to find $\partial V/\partial P$, we differentiate

$$PV - nRT = 0$$

keeping in mind that (besides the constants n and R) the variable T is held fixed:

$$1 \cdot V + P\frac{\partial V}{\partial P} - 0 = 0$$

$$\frac{\partial V}{\partial P} = -\frac{V}{P}$$

To find $\partial V/\partial T$, we differentiate $PV - nRT = 0$ implicitly with respect to T (this time, P is held fixed) and get

$$P\frac{\partial V}{\partial T} - nR = 0$$

$$\frac{\partial V}{\partial T} = \frac{nR}{P}$$

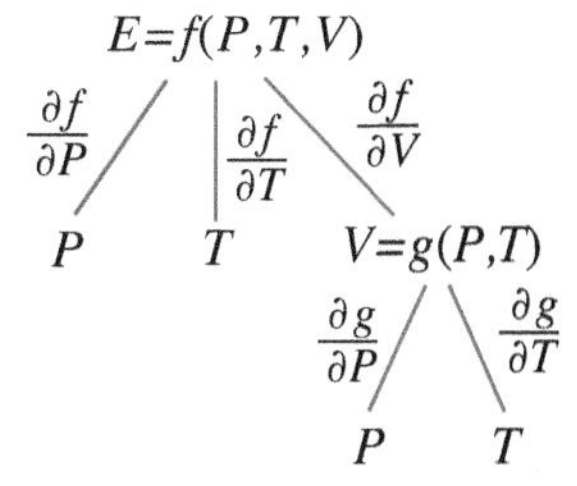

The internal energy E of a gas depends on all three variables from the ideal gas law:

$$E = f(P, T, V)$$

The three variables are not independent: their dependence is given by the equation $PV - nRT = 0$. Thus, if we view V as a function of P and T, i.e., $V = g(P, T)$, then the energy is a function of P and T as well: $E = f(P, T, g(P, T))$.

From the chain rule (see the diagram in Figure 6.6), we get

$$\frac{\partial E}{\partial P} = \frac{\partial f}{\partial P} + \frac{\partial f}{\partial V}\frac{\partial g}{\partial P}$$

$E=f(P,T,V)$; $\frac{\partial f}{\partial P}$ → P; $\frac{\partial f}{\partial T}$ → T; $\frac{\partial f}{\partial V}$ → $V=g(P,T)$; $\frac{\partial g}{\partial P}$ → P; $\frac{\partial g}{\partial T}$ → T

FIGURE 6.6

Tree diagram for computing $\partial E/\partial P$

In this section we used the fact that an equation in two variables implicitly defines a curve in a plane (or in the case of three variables, a surface in space). For the record, we now list explicit conditions that guarantee that this is really so. (Advanced calculus textbooks provide more detail, related proofs, and insight into implicitly defined curves and surfaces.)

Consider an equation $F(x, y) = 0$, and pick a point (a, b) in its domain where $F(a, b) = 0$. Assume the following:

(1) $F(x, y)$ is defined on an open disk $B_r(a, b)$ centred at (a, b); that is, f needs to be defined near (a, b).

(2) The partial derivatives $\partial F/\partial x$ and $\partial F/\partial y$ are continuous on $B_r(a, b)$.

(3) $\partial F/\partial y(a, b) \neq 0$.

Then the equation $F(x, y) = 0$ implicitly defines a function $y = f(x)$, and its derivative is given by Theorem 9.

What we just stated is called the *Implicit Function Theorem*. The theorem can be generalized to include the solutions of $F(x, y, z) = 0$ (i.e., implicitly defined surfaces) and the solutions of systems of equations.

Summary

To calculate the derivative of the composition of functions, we use the **chain rule.** A **tree diagram** shows how a function depends on its variables and helps us write the correct chain rule formula. Using **implicit differentiation,** we can compute the derivative of a function defined **implicitly** by an equation.

6 Exercises

1. Suppose that $z = f(x, y)$, where $x = g(s, t)$ and $y = h(s, t)$. Sketch a tree diagram and find the formulas for $\partial z/\partial s$ and $\partial z/\partial t$.

2. Suppose that $w = g(x, y, z)$, where x, y, and z are functions of t. Sketch a tree diagram and find a formula for $\partial w/\partial t$.

3. Assume that $E = f(M, N)$, where M and N are functions of v. Find a formula for dE/dv.

4. Assume that $W = f(h, m)$, where both h and m depend on v and w. Find the formulas for $\partial W/\partial v$ and $\partial W/\partial w$.

5. Explain how to interpret the hyperbola $y = 1/x$ as an implicitly defined curve.

6. Explain how to interpret the cone $z = \sqrt{x^2 + y^2}$ as an implicitly defined surface.

7–10 ▪ Describe the curve defined implicitly by $F(x,y)=0$.

7. $F(x,y)=x-2y^2-1$
8. $F(x,y)=3x-4y-7$
9. $F(x,y)=x^2y-1$
10. $F(x,y)=ye^x+4$

11. Identify geometrically the set of points given by $F(x,y)=0$, where $F(x,y)=x^2-y^2$.

12. Identify geometrically the set of points given by $F(x,y)=0$, where $F(x,y)=x^2+y^2+1$.

13. Suppose that $z=F(g(x,y),h(x,y))$. Sketch a tree diagram and find the formulas for $\partial z/\partial x$ and $\partial z/\partial y$.

14. Suppose that $z=F(x,g(x,y),h(x,y))$. Sketch a tree diagram and find the formulas for $\partial z/\partial x$ and $\partial z/\partial y$.

15–20 ▪ Find dz/dt using the chain rule. Assume that all functions involved are restricted to the domain where they are differentiable (i.e., satisfy the assumptions of Theorem 8).

15. $z=x^3-2xy+1$, $x=\sin t$, $y=5t$
16. $z=pq-q^{-2}$, $p=\sin t$, $q=\sin 2t$
17. $z=\sqrt{p^2-5q-2}$, $p=t^3$, $q=t$
18. $z=\ln(x^2+y^2+4)$, $x=\sin t$, $y=\cos t$
19. $z=y^2e^{-x}$, $x=t^2$, $y=1/t$
20. $z=x^2y\sin x$, $x=6t$, $y=e^t$

21. The number of whales N depends on the availability of plankton P and the ocean temperature T. Both P and T change with time and thus N is a function of t. Using the chain rule, write a formula for $\partial N/\partial t$ and interpret each derivative involved.

22. Assume that g is a differentiable function such that $g(1)=4$ and $g'(1)=5$. If $f(x,y)=xy^3g(x)$, find $(\partial f/\partial x)(x,y)$, $(\partial f/\partial y)(x,y)$, and $(\partial f/\partial x)(1,1)$.

23. Assume that g is a differentiable function such that $g'(0)=5$. Define $f(x,y)=g(x)+g(x+y^2)$. Find $(\partial f/\partial x)(x,y)$ and $(\partial f/\partial x)(0,0)$.

24. Find $z'(1)$ if $z=x^2-x^3y^{-1}$, $x=3t-4$, and $y=e^{t-1}$.

25. Find $w'(\pi/2)$ if $w=(3x^2-4)e^{y^2}$, $x=\cos t$, and $y=\sin t$.

26. Assume that f is a differentiable function of one variable and $u(x,t)=f(x-5t)$. Express $u_x(x,t)$ and $u_t(x,t)$ in terms of f and f'.

27. Assume that f is a differentiable function of one variable and $u(x,t)=x^2t^3f(x-5t)$. Find $u_x(x,t)$ and $u_t(x,t)$.

28–33 ▪ Find $\partial z/\partial u$ and $\partial z/\partial v$. Assume that all functions involved are restricted to the domain where they are differentiable (so that the chain rule formula applies).

28. $z=y^2e^{-x}$, $x=2u-5v$, $y=-u-4v$
29. $z=\ln(a+b^3-2)$, $a=uv$, $b=u^2-v^2$
30. $z=\dfrac{ab-1}{b^2+1}$, $a=3u$, $b=uv$.
31. $z=\sin(xy)$, $x=u^2v$, $y=-2uv^4$
32. $z=\arctan(y/x)$, $x=u/v$, $y=v/u^2$
33. $z=\dfrac{x^2-y}{1-xy}$, $x=2uv$, $y=5v$

34. Find $z'(t)$ if $z=e^{xy}-x^2$, where $x=4t^2$ and $y=7$.

35–40 ▪ Find dy/dx by implicit differentiation.

35. $x^3+y^2=1$
36. $3-xy=\sin x$
37. $e^{xy}-y^2=1$
38. $\sqrt{x^2+y^2}=1-x$
39. $\sin(2x-y)=\cos(x-3y)$
40. $\ln(2x+y^3)=\ln 2+x$

41–44 ▪ Use implicit differentiation to find $\partial z/\partial x$ and $\partial z/\partial y$.

41. $xyz = 64$

42. $x^2y^3 + \ln z = 4$

43. $e^{xz} - eyz = 0$

44. $x^{-1}y^{-2}z^{-3} = 1$

45. In the text, we mentioned that solving $(x^2 + y^2)^2 - x^2 + y^2 = 0$ explicitly for y is difficult. Not really: simplify the equation first to get $y^4 + (2x^2 + 1)y^2 + x^4 - x^2 = 0$ and then use the quadratic formula to obtain the solutions $y = \pm\frac{1}{2}\sqrt{-2 - 4x^2 + 2\sqrt{8x^2 + 1}}$.

46. Find a differentiable function $F(x, y)$ such that the set $F(x, y) = 0$ consists of the single point $(2, 3)$.

47. Find a differentiable function $F(x, y)$ such that the set $F(x, y) = 0$ consists of two parallel lines.

7 Second-Order Partial Derivatives and Applications

By calculating the partial derivatives of the partial derivatives, we obtain the **second-order partial derivatives.** We use numerical and geometric approaches to reason about and understand these derivatives. As a major application, we construct the **degree-2 Taylor polynomial.** The corresponding approximation, called the **quadratic approximation,** is an improvement on the linear approximation we studied earlier.

Second-Order Partial Derivatives

The partial derivatives $\partial f/\partial x$ and $\partial f/\partial y$ of a function $f(x, y)$ of two variables are functions of the same variables. By differentiating them we obtain *second-order partial derivatives.*

Differentiating $\partial f/\partial x$ with respect to x, we obtain the second-order derivative

$$\frac{\partial}{\partial x}\left(\frac{\partial f}{\partial x}\right) = \frac{\partial^2 f}{\partial x^2}$$

Using subscripts to denote partial derivatives, we write

$$(f_x)_x = f_{xx}$$

Differentiating $\partial f/\partial x$ with respect to y, we obtain

$$\frac{\partial}{\partial y}\left(\frac{\partial f}{\partial x}\right) = \frac{\partial^2 f}{\partial y \partial x} \quad \text{or} \quad (f_x)_y = f_{xy}$$

As well,

$$\frac{\partial}{\partial x}\left(\frac{\partial f}{\partial y}\right) = \frac{\partial^2 f}{\partial x \partial y} \quad \text{or} \quad (f_y)_x = f_{yx}$$

and

$$\frac{\partial}{\partial y}\left(\frac{\partial f}{\partial y}\right) = \frac{\partial^2 f}{\partial y^2} \quad \text{or} \quad (f_y)_y = f_{yy}$$

Why do we need second-order derivatives?

There are several reasons. First, they give information on how (the first-order) partial derivatives behave, i.e., how the rates of change change themselves. For instance, consider a population of wolves $W(F, D)$ that depends on the availability of food F and the proximity D of urban areas. From $\partial W/\partial F > 0$ we conclude that with an increase in the availability of food (keeping the distance from urban areas unchanged), the wolf population increases. If, in addition

$$\frac{\partial}{\partial F}\left(\frac{\partial W}{\partial F}\right) = \frac{\partial^2 W}{\partial F^2} > 0$$

then the wolf population is increasing at an increasing rate.

Second-order partial derivatives help us construct the quadratic approximation of a function, which is an improvement on the linear approximation that we discussed earlier.

Using partial derivatives, we build partial differential equations, which are an essential tool in many applications in life sciences and elsewhere.

To start, we learn how to calculate second-order partial derivatives.

Example 7.1 Computing Second-Order Partial Derivatives

Compute all second-order partial derivatives for $f(x,y) = x^3 e^{y^2} + 2y$.

From $f_x(x,y) = 3x^2 e^{y^2}$, we compute

$$f_{xx}(x,y) = 6xe^{y^2}$$
$$f_{xy}(x,y) = 3x^2 e^{y^2}(2y) = 6x^2 ye^{y^2}$$

From $f_y(x,y) = 2x^3 ye^{y^2} + 2$, we compute

$$f_{yx}(x,y) = 6x^2 ye^{y^2}$$
$$f_{yy}(x,y) = 2x^3(e^{y^2} + ye^{y^2}(2y)) = 2x^3 e^{y^2}(1 + 2y^2)$$

Example 7.2 Level Curves and Second-Order Partial Derivatives

Determine the signs of the second-order partial derivatives f_{xx}, f_{yy}, g_{xx}, and g_{yy} at A of the functions f and g whose level curves are given in Figure 7.1.

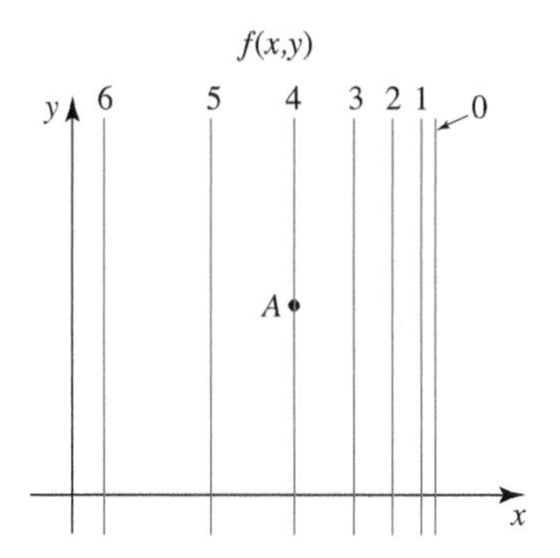

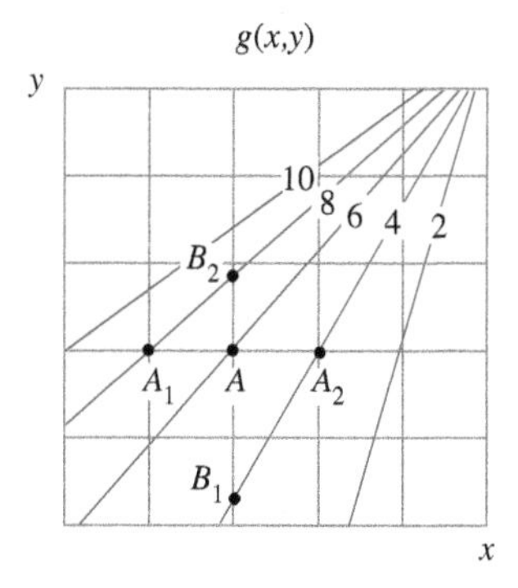

FIGURE 7.1
Level curves of functions f and g

Moving horizontally in the direction of the x-axis, we see that f decreases as we pass through A, so $f_x(A) < 0$. Moreover, we meet the level curves more and more quickly (i.e., the change of -1 in the values of f happens over shorter and shorter distances); from this observation, we conclude that $f_{xx}(A) < 0$.

Looking at the level curves, we see that f does not change in the vertical directions (i.e., for a fixed value of x, f is a constant function); thus, $f_y(A) = 0$. Note that f_y is zero not just at A, but at all points on the vertical line through A. In other words, as x is kept fixed, f_y (as function of y) is a constant function. Thus, $f_{yy}(A) = 0$.

The function g is decreasing in the x direction at A, so $g_x(A) < 0$. Moreover, the level curves indicate that g decreases by the same amount (that is, g_x has the same value at A, A_1, and A_2) and thus $g_{xx}(A) = 0$.

In the y direction, g increases at A, and so $g_y(A) > 0$ (note that $g_y(B_1) > 0$ and $g_y(B_2) > 0$ as well). As we move vertically in the direction of the y-axis, we notice that the same change in y (2 units) occurs over shorter and shorter distances; thus, g_y increases at an increasing rate and $g_{yy}(A) > 0$.

We have one more question related to the level curves of $g(x,y)$ in Figure 7.1: what is the sign of g_{xy} at A; i.e., how does g_x change in the vertical direction? We see that $g_x(A) < 0$ and $g_x(B_2) < 0$; we need to figure out how the two numbers compare. Note that the same change in the function (-2) occurs more quickly at B_2 than at A; thus, $g_x(B_2) < g_x(A)$, which means that g_x is decreasing in the y direction at A, and so $g_{xy} < 0$.

Example 7.3 Using a Table of Values to Estimate Partial Derivatives

Using the table of values of the function $f(x, y)$ given in Table 7.1, estimate $f_x(1, 0)$, $f_{xx}(1, 0)$, and $f_{xy}(1, 0)$.

Table 7.1

	$x = 0.9$	$x = 1$	$x = 1.1$	$x = 1.2$
$y = -0.1$	5.9	6.1	6.8	6.7
$y = 0$	5.6	6	6.2	6.3
$y = 0.1$	5.4	5.7	6.1	6.5

To estimate $f_x(1, 0)$, we use $(1, 0)$ and the nearby point $(1.1, 0)$:

$$f_x(1,0) \approx \frac{f(1.1,0) - f(1,0)}{0.1} = \frac{6.2 - 6}{0.1} = 2$$

To estimate $f_{xx}(1, 0)$, we need the value of f_x at a nearby point (in the horizontal direction). So we estimate

$$f_x(1.1,0) \approx \frac{f(1.2,0) - f(1.1,0)}{0.1} = \frac{6.3 - 6.2}{0.1} = 1$$

Thus,

$$f_{xx}(1,0) \approx \frac{f_x(1.1,0) - f_x(1,0)}{0.1} = \frac{1 - 2}{0.1} = -10$$

To estimate $f_{xy}(1, 0)$, we need the value of f_x at a nearby point (in the vertical direction). From

$$f_x(1,0.1) \approx \frac{f(1.1,0.1) - f(1,0.1)}{0.1} = \frac{6.1 - 5.7}{0.1} = 4$$

we get

$$f_{xy}(1,0) \approx \frac{f_x(1,0.1) - f_x(1,0)}{0.1} = \frac{4 - 2}{0.1} = 20$$

To estimate $f_{yx}(1, 0)$, we use

$$f_{yx}(1,0) \approx \frac{f_y(1.1,0) - f_y(1,0)}{0.1}$$

From

$$f_y(1,0) \approx \frac{f(1,0.1) - f(1,0)}{0.1} = \frac{5.7 - 6}{0.1} = -3$$

and

$$f_y(1.1,0) \approx \frac{f(1.1,0.1) - f(1.1,0)}{0.1} = \frac{6.1 - 6.2}{0.1} = -1$$

we get

$$f_{yx}(1,0) \approx \frac{f_y(1.1,0) - f_y(1,0)}{0.1} = \frac{-1 - (-3)}{0.1} = 20$$

This numerical example, as well as Example 7.1, suggests that the second-order partial derivatives (also called the *mixed partial derivatives*) f_{xy} and f_{yx} might be equal. Although not true in general, the equality does hold under certain conditions, stated in the following theorem.

Theorem 11 Equality of Mixed Partial Derivatives

Assume that a function f is defined in an open disk $B_r(a, b)$ around a point (a, b) and that the partial derivatives f_{xy} and f_{yx} are continuous on $B_r(a, b)$. Then $f_{xy}(a, b) = f_{yx}(a, b)$.

Good news: all functions that we'll meet in this book will satisfy the assumptions of this theorem.

Quadratic Approximation

A differentiable function $y = f(x)$ can be approximated by a linear function (called the *linearization* at $x = a$, or the *Taylor polynomial of degree 1 at* $x = a$)

$$L_a(x) = f(a) + f'(a)(x - a)$$

A better approximation of f near $x = a$ is provided by the quadratic function

$$T_2(x) = f(a) + f'(a)(x - a) + \frac{f''(a)}{2}(x - a)^2 \qquad (7.1)$$

known as the *quadratic approximation at* $x = a$ or the *Taylor polynomial of degree 2 at* $x = a$.

In Section 5 we constructed the linearization of a function $f(x, y)$ near a point (a, b) in its domain. More precisely, we have the following theorem:

Theorem 12 Linearization of a Function of Two Variables

Assume that f has continuous partial derivatives $\partial f/\partial x$ and $\partial f/\partial y$ near (a, b). Then

$$f(x, y) \approx L_{(a,b)}(x, y)$$

for (x, y) near (a, b), where

$$L_{(a,b)}(x, y) = f(a, b) + f_x(a, b)(x - a) + f_y(a, b)(y - b)$$

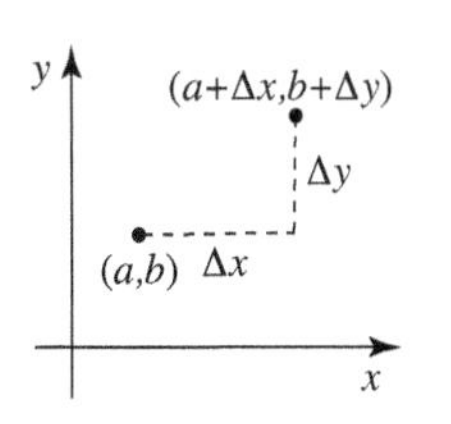

FIGURE 7.2
Deriving the quadratic approximation

To improve this approximation, we now construct a quadratic approximation for f. In order to do so, we approximate the value $f(a + \Delta x, b + \Delta y)$ of $f(x, y)$ at a point $(a + \Delta x, b + \Delta y)$ near (a, b); see Figure 7.2.

The trick we use is to reduce our problem to the one-variable case and use the quadratic approximation (7.1). We consider the function

$$F(t) = f(a + t\Delta x, b + t\Delta y) \qquad (7.2)$$

of *one* variable t (all of a, b, Δx, and Δy are viewed as constants). Note that $F(0) = f(a, b)$ and $F(1) = f(a + \Delta x, b + \Delta y)$, which is the value that we are interested in. The degree-2 Taylor polynomial for F at $t = 0$ is

$$T_2(t) = F(0) + F'(0)t + \frac{F''(0)}{2}t^2 \qquad (7.3)$$

Once we figure it out, we will substitute $t = 1$ into (7.3) and obtain the approximation

$$f(a + \Delta x, b + \Delta y) = F(1) \approx T_2(1) = F(0) + F'(0) + \frac{F''(0)}{2} \qquad (7.4)$$

Now we calculate $F'(0)$ and $F''(0)$. Differentiating (7.2) using the chain rule, we have

$$F'(t) = f_x(a + t\Delta x, b + t\Delta y)\Delta x + f_y(a + t\Delta x, b + t\Delta y)\Delta y$$

and thus

$$F'(0) = f_x(a, b)\Delta x + f_y(a, b)\Delta y$$

To compute $F''(t)$, we use the chain rule again:

$$\begin{aligned} F''(t) = {} & \big[f_{xx}(a + t\Delta x, b + t\Delta y)\Delta x + f_{xy}(a + t\Delta x, b + t\Delta y)\Delta y\big]\Delta x \\ & + \big[f_{yx}(a + t\Delta x, b + t\Delta y)\Delta x + f_{yy}(a + t\Delta x, b + t\Delta y)\Delta y\big]\Delta y \end{aligned}$$

and

$$F''(0) = \big[f_{xx}(a,b)\Delta x + f_{xy}(a,b)\Delta y\big]\Delta x + \big[f_{yx}(a,b)\Delta x + f_{yy}(a,b)\Delta y\big]\Delta y$$
$$= f_{xx}(a,b)\Delta x^2 + 2f_{xy}(a,b)\Delta x\Delta y + f_{yy}(a,b)\Delta y^2$$

assuming that the mixed partial derivatives of f are equal ($f_{xy} = f_{yx}$; see (Theorem 11).

Substituting $F(0)$, $F'(0)$, and $F''(0)$ into (7.4), we obtain the polynomial

$$T_2(x,y) = f(a,b) + f_x(a,b)\Delta x + f_y(a,b)\Delta y$$
$$+ \frac{1}{2}\left(f_{xx}(a,b)\Delta x^2 + 2f_{xy}(a,b)\Delta x\Delta y + f_{yy}(a,b)\Delta y^2\right)$$

Rewriting this formula, keeping in mind that $\Delta x = x - a$ and $\Delta y = y - b$, we obtain the following result.

Theorem 13 Quadratic Approximation and Taylor Polynomial

Assume that all second-order partial derivatives of a function $f(x,y)$ are continuous near (a,b). For (x,y) near (a,b), the *quadratic approximation of f* is given by

$$f(x,y) \approx T_2(x,y),$$

where T_2 is the *degree-2 Taylor polynomial*

$$T_2(x,y) = f(a,b) + f_x(a,b)(x-a) + f_y(a,b)(y-b)$$
$$+ \frac{1}{2}\left(f_{xx}(a,b)(x-a)^2 + 2f_{xy}(a,b)(x-a)(y-b) + f_{yy}(a,b)(y-b)^2\right)$$

In order to avoid using messy notation, we do not add (a,b) to the notation for the Taylor polynomial $T_2(x,y)$, but we will always make sure that we clearly identify what the base point (a,b) is.

Note that the linear part of $T_2(x,y)$ is the linearization $L_{(a,b)}(x,y)$.

Example 7.4 Calculating a Degree-2 Taylor Polynomial

Compute the degree-2 Taylor polynomial for the function $f(x,y) = \sin(x+y) + \cos(x-2y)$ based at the point $(0,0)$.

▶ The degree-2 Taylor polynomial is given by

$$T_2(x,y) = f(0,0) + f_x(0,0)(x-0) + f_y(0,0)(y-0)$$
$$+ \frac{1}{2}\left(f_{xx}(0,0)(x-0)^2 + 2f_{xy}(0,0)(x-0)(y-0) + f_{yy}(0,0)(y-0)^2\right)$$

We organize the calculations of the values of the function and its partial derivatives in Table 7.2.

Table 7.2

f and its partial derivatives	Value at $(0,0)$
$f = \sin(x+y) + \cos(x-2y)$	1
$f_x = \cos(x+y) - \sin(x-2y)$	1
$f_y = \cos(x+y) + 2\sin(x-2y)$	1
$f_{xx} = -\sin(x+y) - \cos(x-2y)$	−1
$f_{xy} = -\sin(x+y) + 2\cos(x-2y)$	2
$f_{yx} = -\sin(x+y) + 2\cos(x-2y)$	2
$f_{yy} = -\sin(x+y) - 4\cos(x-2y)$	−4

Thus,

$$T_2(x,y) = 1 + (1)x + (1)y + \frac{1}{2}((-1)x^2 + 2(2)xy + (-4)y^2)$$
$$= 1 + x + y - \frac{1}{2}x^2 + 2xy - 2y^2$$

The first three terms of $T_2(x, y)$ belong to the linearization

$$L_{(0,0)}(x, y) = 1 + x + y$$

of f at the origin. To compare the two approximations, we draw the level curves of f (Figure 7.3a), its linearization $L_{(0,0)}$ (Figure 7.3b), and its degree-2 Taylor polynomial T_2 (Figure 7.3c).

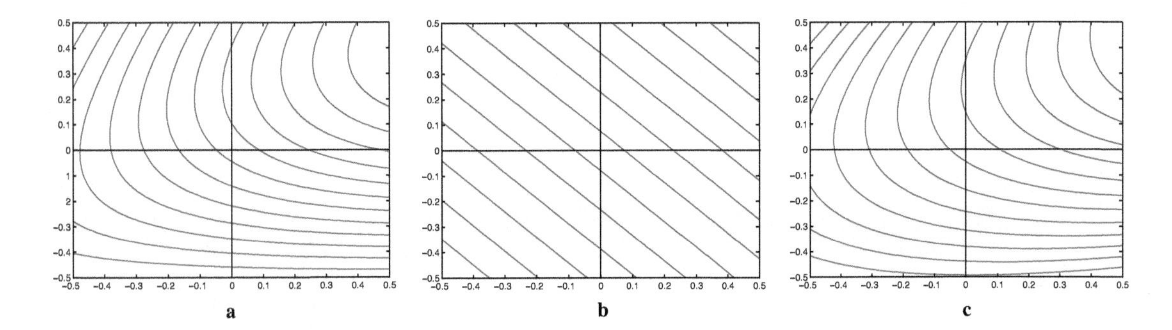

FIGURE 7.3

Level curves of f, its linearization, and its degree-2 Taylor polynomial

An alternative to calculating partial derivatives in Example 7.4 consists of using Taylor polynomials in one variable, and discarding all terms beyond the quadratic. Recalling that $\sin A \approx A$ and $\cos A \approx 1 - A^2/2$ for A near 0, we write

$$\sin(x+y) + \cos(x-2y) \approx (x+y) + 1 - \frac{1}{2}(x-2y)^2$$
$$= x + y + 1 - \frac{1}{2}\left(x^2 - 4xy + 4y^2\right)$$
$$= 1 + x + y - \frac{1}{2}x^2 + 2xy - 2y^2$$

Example 7.5 Calculating the Linearization and the Degree-2 Taylor Polynomial

Find the linearization and the degree-2 Taylor polynomial for $f(x, y) = xye^{-x^2} + y^2$ at $(0, 2)$.

As in Example 7.4, we organize our calculations in a table (see Table 7.3).

Table 7.3

f and its partial derivatives	At $(0, 2)$
$f = xye^{-x^2} + y^2$	4
$f_x = ye^{-x^2} + xye^{-x^2}(-2x) = ye^{-x^2} - 2x^2ye^{-x^2}$	2
$f_y = xe^{-x^2} + 2y$	4
$f_{xx} = ye^{-x^2}(-2x) - 4xye^{-x^2} - 2x^2ye^{-x^2}(-2x) = -6xye^{-x^2} + 4x^3ye^{-x^2}$	0
$f_{xy} = e^{-x^2} - 2x^2e^{-x^2}$	1
$f_{yx} = e^{-x^2} + e^{-x^2}(-2x) = e^{-x^2} - 2x^2e^{-x^2}$	1
$f_{yy} = 2$	2

We find

$$\begin{aligned}L_{(0,2)}(x,y) &= f(0,2) + f_x(0,2)(x-0) + f_y(0,2)(y-2)\\ &= 4 + 2x + 4(y-2)\\ &= -4 + 2x + 4y\end{aligned}$$

and

$$\begin{aligned}T_2(x,y) &= f(0,2) + f_x(0,2)(x-0) + f_y(0,2)(y-2)\\ &\quad + \frac{1}{2}\left(f_{xx}(0,2)(x-0)^2 + 2f_{xy}(0,2)(x-0)(y-2) + f_{yy}(0,2)(y-2)^2\right)\\ &= 4 + 2x + 4(y-2) + \frac{1}{2}\left((0)x^2 + 2x(y-2) + 2(y-2)^2\right)\\ &= 4 + 2x + 4y - 8 + xy - 2x + y^2 - 4y + 4\\ &= xy + y^2\end{aligned}$$

In Table 7.4 we compare the accuracy of the linear and the quadratic approximations of f at several points near $(0,2)$.

Table 7.4

(x,y)	Distance from $(0,2)$	$L_{(0,2)}(x,y)$	$T_2(x,y)$	$f(x,y)$
$(0.1, 1.9)$	0.141	3.80	3.80	3.798109
$(0.1, 1.95)$	0.112	4	3.9975	3.995560
$(-0.05, 2.01)$	0.051	3.94	3.9396	3.939851
$(0.01, 1.999)$	0.01	4.016	4.015991	4.015989

How does

$$\begin{aligned}T_2(x,y) &= f(a,b) + f_x(a,b)(x-a) + f_y(a,b)(y-b)\\ &\quad + \frac{1}{2}\left(f_{xx}(a,b)(x-a)^2 + 2f_{xy}(a,b)(x-a)(y-b) + f_{yy}(a,b)(y-b)^2\right)\end{aligned}$$

relate to $f(x,y)$ algebraically?

We know that the two have the same values at the base point: $T_2(a,b) = f(a,b)$. From

$$\begin{aligned}\frac{\partial T_2}{\partial x} &= 0 + f_x(a,b)(1) + 0\\ &\quad + \frac{1}{2}\left(f_{xx}(a,b)\cdot 2(x-a) + 2f_{xy}(a,b)(1)(y-b) + 0\right)\\ &= f_x(a,b) + f_{xx}(a,b)(x-a) + f_{xy}(a,b)(y-b)\end{aligned}$$

we get that $\partial T_2/\partial x(a,b) = f_x(a,b)$. Differentiating again,

$$\frac{\partial^2 T_2}{\partial x^2} = 0 + f_{xx}(a,b) + 0 = f_{xx}(a,b)$$

and, at (a,b), $\partial^2 T_2/\partial x^2(a,b) = f_{xx}(a,b)$. In the same way, we check the remaining partial derivatives.

In conclusion, if T_2 is the degree-2 Taylor polynimial of f at (a,b), then $T_2(a,b) = f(a,b)$; as well, all respective first-order and second-order partial derivatives of T_2 and f agree at (a,b).

Summary The **second-order partial derivatives** are partial derivatives of partial derivatives. A function of two variables has four second-order partial derivatives. Under certain conditions, the two **mixed partial derivatives** are equal. The **degree-2 Taylor polynomial** is an improvement over the linearization in terms of approximating the values of a function.

7 Exercises

1. Write the limit definition of $f_{yy}(2,7)$ as $(f_y)_y$.

2. Write the limit definition of $f_{xy}(a,b)$ as $(f_x)_y$.

3. Sketch a contour diagram of a function for which $f_{xx} > 0$ at all points in the plane.

4. Sketch a contour diagram of a function for which $f_x > 0$ and $f_{xx} < 0$ at all points in the plane.

5. Can $g(x,y) = 2 - x - x^2 - xy^2 + y$ be a degree-2 Taylor polynomial of some function? If so, find that function.

6. Give an example of a function whose degree-2 Taylor polynomial at $(0,0)$ is $3 + y - 2x^2$.

7. Give an example of a function whose degree-2 Taylor polynomial at the point $(2,1)$ is $2 + (x-2)^2 + (x-2)(y-1)$.

8–9 ▪ Looking at the contour diagram of each function $f(x,y)$, determine whether the partial derivatives $f_x(A)$, $f_y(A)$, $f_{xx}(A)$, $f_{xy}(A)$, and $f_{yy}(A)$ are positive, negative, or zero.

8.

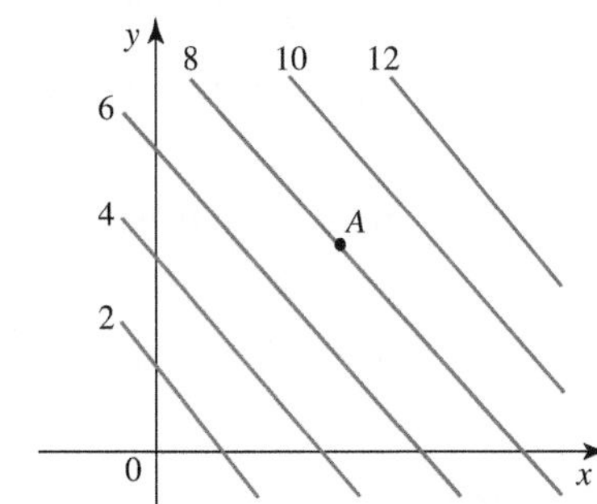

9.

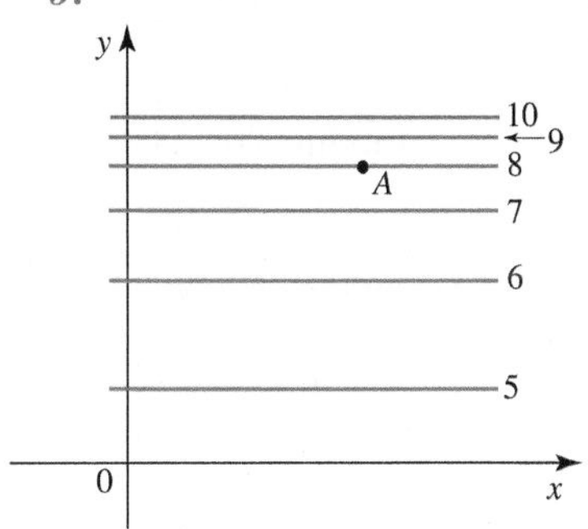

10–13 ▪ Using the values of the function given in Table 7.5, estimate each partial derivative.

Table 7.5

	$x=3$	$x=4$	$x=5$	$x=6$
$y=0$	2.3	2.8	2.9	2.6
$y=1$	3.2	3.5	3.1	2.8
$y=2$	3.2	3.3	3.4	3.3

10. $f_x(4,1)$ and $f_{xx}(4,1)$

11. $f_{xy}(4,1)$

12. $f_y(4,1)$ and $f_{yy}(4,1)$

13. $f_y(5,1)$ and $f_{yx}(5,1)$

14. Assume that f is a differentiable function of one variable and $u(x,t) = f(2x - 3t)$. Express $u_{xx}(x,t)$, $u_{tx}(x,t)$, and $u_{tt}(x,t)$ in terms of the derivatives of f.

15. Assume that f is a differentiable function of one variable and define $u(x,t) = f(x - st)$, where s is a constant. Express $u_{xx}(x,t)$, $u_{tx}(x,t)$, and $u_{tt}(x,t)$ in terms of the derivatives of f.

16–23 ▪ Find the indicated partial derivatives.

16. $z = e^{xy}$; z_{xx}, z_{yx}, z_{yy}

17. $z = \sqrt{x^2 + y^2}$; z_{xx}, z_{xy}, z_{yy}

18. $f(x,y) = \dfrac{xy}{x^2+1}$; f_{xx}, f_{yx}, f_{yy}

19. $f(x,y) = 3 - x$; f_{xx}, f_{yx}, f_{yy}

20. $f(x, y) = x \ln y$; f_{xx}, f_{yy}

21. $g(x, y) = yx^{-1}$; g_{xx}, g_{xy}, g_{yx}, g_{yy}

22. $f(x, y) = \sin x \sin y$; f_{xx}, f_{yx}

23. $z = (1 - xy)^{-1}$; z_{xx}, z_{xy}

24. Give an example of a function whose degree-2 Taylor polynomial at $(1, 0)$ is x^2.

25. Give an example of two different functions f and g whose degree-2 Taylor polynomials at $(0, 0)$ are equal.

26–33 ▪ Find the degree-2 Taylor polynomial of the given function at the given point.

26. $f(x, y) = e^{-x^2-y^2}$; $(0, 0)$

27. $f(x, y) = 1 - x - x^2 + 3y^2$; $(0, 0)$

28. $f(x, y) = 1 - x - x^2 + 3y^2$; $(1, 2)$

29. $f(x, y) = \sin(2x - y)$; $(0, 0)$

30. $f(x, y) = \sin(2x - y)$; $(\pi/2, 0)$

31. $f(x, y) = e^{-x} \sin y$; $(1, 0)$

32. $f(x, y) = (xy)^{-1}$; $(2, 3)$

33. $f(x, y) = \ln(1 - x^2 - y^2)$; $(0, 0)$

34. Find the linearization and the degree-2 Taylor polynomial of the function $f(x, y) = x^2 \arctan y$ at $(1, 0)$. Compare the two approximations of $f(1.05, 0.05)$ with the value of the function.

35. Find the linearization and the degree-2 Taylor polynomial of the function $f(x, y) = \sqrt{3x + y - 1}$ at $(1, -1)$. Compare the two approximations of $f(0.9, -1)$ with the value of the function.

36. Compute the linear and the quadratic approximations of $f(x, y) = (xy)^{-1}$ at $(1, 1)$. Compare the values of the two approximations at $(1.1, 0.9)$ with the value $f(1.1, 0.9)$.

37. Find the degree-2 Taylor polynomial of the function $f(x, y) = x \sin y$ at $(1, 0)$ and use it to draw an approximation of the contour diagram of $f(x, y)$ near $(1, 0)$.

38. Find the degree-2 Taylor polynomial of the function $f(x, y) = y \cos x^2$ at $(0, 0)$ and use it to draw an approximation of the contour diagram of $f(x, y)$ near $(0, 0)$.

39–42 ▪ Use the degree-2 Taylor polynomial to approximate each expression.

39. $1.01 \ln 1.08$

40. $3.99 \arctan 0.1$

41. $e^{0.9^2-0.05^2}$

42. $\sin 0.9 \cos 0.1$

43. How many different second-order partial derivatives does a function of 5 variables have? Of 10 variables? Of 100 variables?

8 Partial Differential Equations

The reason why we define derivatives—ordinary (in the case of functions of one variable) and partial (for functions of several variables)—is to **study change.** Most often, we do this by analyzing (and solving) an **initial value problem** (IVP).

An IVP consists of a differential equation and an initial value. A differential equation establishes a relationship between the rate of change of a function and the function itself and/or its independent variable(s). The initial value represents the measurement of the quantities involved at the start of the investigation or experiment.

Now that we have learned about partial derivatives and know how to interpret them, we build several **partial differential equations** that are commonly found in applications in life sciences and elsewhere.

Travelling Waves

Assume that the function $u(x,t)$ describes the height of a sea wave at time t, at a distance of x units from our point of reference (which, for convenience, we take to be the origin). The height is measured with respect to the level of the still sea (thus, if there are no waves at time t, $u(x,t) = 0$ at all locations x).

Assume that the wave is moving horizontally with a constant speed of s units in the direction of the positive x-axis. In Figure 8.1a we show a snapshot of the wave taken at the time t. The crest of the wave is at the location x_2. At x_1, the water level is decreasing (i.e., $u(x,t)$ is a decreasing function of t), and at x_3 it is increasing.

We make one more assumption: the wave does not change its shape over time. A wave with this property is called a *travelling wave.* In Figure 8.1b we show a travelling wave at three different times; all positions are horizontal shifts of each other. Travelling waves occur when the motion is not confined to a given space (unlike, for instance, water waves in a bathtub).

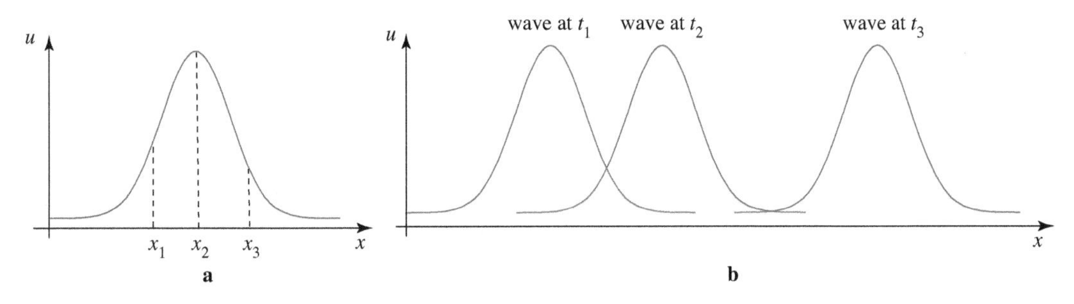

FIGURE 8.1

Snapshot of a wave, and a wave moving

A travelling wave does not have to be a water wave. Travelling waves have been identified in a number of phenomena, including propagation of nerve signals within neurons, the spread of genes in a population, and the spread of colonies of micro-organisms.

How can we describe a travelling wave using the partial derivatives of u? We start by checking the signs of the partial derivatives u_x and u_t at the points x_1, x_2, and x_3 shown in Figure 8.1a.

The partial derivative $u_x(x,t)$ is the slope of the tangent to the curve $u(x,t)$ at x (t is fixed). At x_1, $u(x,t)$ is an increasing function of x and so $u_x(x_1,t) > 0$.

At x_2, the tangent to the graph $u(x,t)$ is horizontal; thus, $u_x(x_1,t) = 0$. Since $u(x,t)$ is decreasing at x_3, we conclude that $u_x(x_3,t) < 0$.

Now fix a location x. The function $u(x,t)$ describes the height of the water at that fixed location. Since the crest of the wave has passed the point x_1, the height of the water is decreasing, and so $u_t(x_1,t) < 0$. The crest is approaching x_3, and the water level is rising; thus $u_t(x_3,t) > 0$. At x_2, the wave height has reached its maximum at the time t, i.e., $u_t(x_2,t) = 0$.

To summarize our analysis: either $u_t(x,t)$ and $u_x(x,t)$ are of opposite signs, or they are both zero. This observation indicates that they might be related. We now show that this is indeed so: the two partial derivatives are actually proportional, i.e.,

$$u_x(x,t) = Au_t(x,t)$$

for some constant $A \neq 0$.

Figure 8.2 shows the position of a wave at time t and a short time $t+\Delta t$ later. We now compute u_x at the point A.

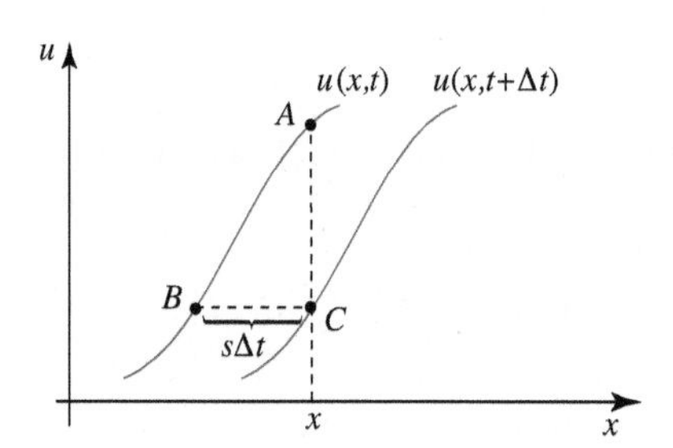

FIGURE 8.2

Figuring out the relationship between the partial derivatives of u

The derivative $u_x(x,t)$ is the slope of the tangent to the curve $u(x,t)$ at A. We will approximate this slope by the slope of the secant line connecting points A and B.

What are the coordinates of B? Note that B lies on the curve $u(x,t)$, so its time coordinate is t. Its distance from C is equal to the distance the wave travels in time Δt; since the speed of the wave is s, that distance is (speed times time) $s\Delta t$. Thus, the coordinates of B are $(x - s\Delta t, t)$.

The length AC is the difference (at a fixed location x) in the height of the wave at time t and the height Δt time units later, that is, $u(x,t) - u(x,t+\Delta t)$. So the slope of the secant is

$$\text{slope of } AB = \frac{AC}{BC} = \frac{u(x,t) - u(x,t+\Delta t)}{s\Delta t}$$

It follows that

$$\begin{aligned} u_x(x,t) &\approx \text{slope of } AB = \frac{u(x,t) - u(x,t+\Delta t)}{s\Delta t} \\ &= -\frac{1}{s}\frac{u(x,t+\Delta t) - u(x,t)}{\Delta t} \\ &\approx -\frac{1}{s}u_t(x,t) \end{aligned}$$

and so the travelling wave satisfies the partial differential equation

$$u_t(x,t) = -su_x(x,t) \tag{8.1}$$

Assume that the function $y = f(x)$ represents the shape of a travelling wave (Figure 8.3a) and that it is moving at a constant speed of s units.

In t units of time, the wave travels the distance st; since it keeps its shape, at time t it must be the same as the graph of $y = f(x)$ shifted st units to the right, i.e., its graph must be given by $y = f(x - st)$; see Figure 8.3b.

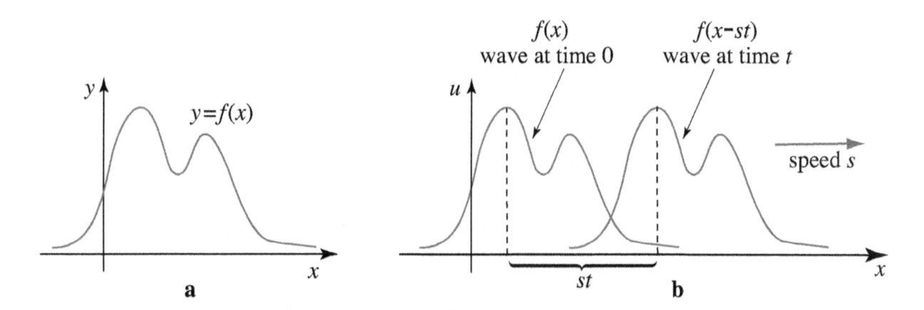

FIGURE 8.3

Travelling wave $u(x,t) = f(x - st)$

In this way, we have obtained the function $u(x,t) = f(x - st)$ that generates the travelling wave for a given shape $y = f(x)$.

Example 8.1 Travelling Wave

Using a function $u(x,t)$ of two variables, describe the wave with shape at $t = 0$ given by $f(x) = \sin 2x$, travelling at a speed of 25 km/h in the direction of the positive x-axis. Check that $u(x,t)$ satisfies the partial differential equation (8.1).

▶ The speed is $s = 25$ and so the wave is given by

$$u(x,t) = f(x - 25t) = \sin 2(x - 25t) = \sin(2x - 50t)$$

The partial derivatives of u are

$$u_x(x,t) = 2\cos(2x - 50t)$$

and

$$u_t(x,t) = -50\cos(2x - 50t)$$

It follows that

$$u_t(x,t) = -50\cos(2x - 50t) = -25\big[2\cos(2x - 50t)\big] = -25u_x(x,t)$$

which is (8.1) with $s = 25$. ▲

Example 8.2 Travelling Wave

Assume that the wave has the shape of the graph $f(x) = e^{-x^2}$ at time $t = 0$. The travelling wave of speed 3 units generated by $f(x)$ has equation

$$u(x,t) = f(x - 3t) = e^{-(x-3t)^2}$$

From

$$u_x(x,t) = e^{-(x-3t)^2}(-2(x - 3t)) = -2(x - 3t)e^{-(x-3t)^2}$$

and

$$u_t(x,t) = e^{-(x-3t)^2}(-2(x - 3t)(-3)) = 6(x - 3t)e^{-(x-3t)^2}$$

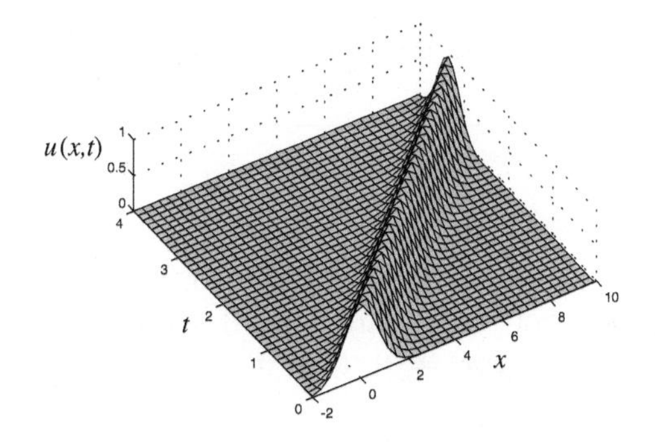

FIGURE 8.4

Travelling wave $u(x,t) = e^{-(x-3t)^2}$

we see that

$$u_t(x,t) = 6(x-3t)e^{-(x-3t)^2} = -3\left[-2(x-3t)e^{-(x-3t)^2}\right] = -3u_x(x,t)$$

so u indeed satisfies the partial differential equation (8.1). The plot of the function $u(x,t)$ in Figure 8.4 shows how the wave moves. ▲

Next, we study a different type of wave.

One-Dimensional Wave Equation

Consider the motion of a vibrating string fixed at both ends (such as a guitar string). By $u(x,t)$ we denote the vertical displacement of the string at the location x at time t. The displacement is measured from the equilibrium position (x-axis); see Figure 8.5.

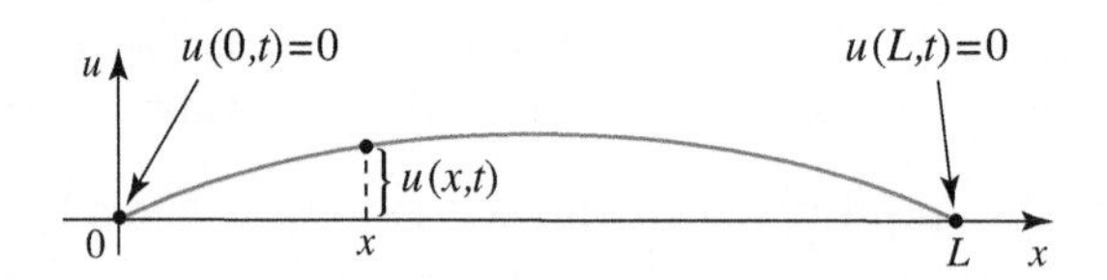

FIGURE 8.5

A snapshot of a vibrating string

The ends of the string are fixed at $x = 0$ and $x = L$. "Fixed" means that they do not move; i.e., $u(0,t) = 0$ and $u(L,t) = 0$ for all time t.

It can be shown that $u(x,t)$ satisfies the *wave equation*

$$u_{tt}(x,t) = a^2 u_{xx}(x,t) \tag{8.2}$$

where the constant a incorporates the properties of the string (such as density and tension). The square indicates that the two second partials have to have the same sign (or are simultaneously zero); we'll get back to this soon.

Partial differential equation (8.2) describes vibrations in one dimension (up–down) and is used to model a wide variety of phenomena (sea waves, sound waves, elasticity, transversal motion of the wall of a blood vessel caused by blood movement, etc.). We can generalize the wave equation to describe two-dimensional vibrations (such as motions of all kinds of membranes) or three-dimensional vibrations (such as the spread of sound waves through space or waves of low or high pressure moving in Earth's atmosphere). We will not derive equation (8.2), but we will be able to convince ourselves that it makes sense.

The term $u_{xx}(x,t)$ describes the concavity of the string (at a fixed time t). The function $u(x,t)$, for fixed x, represents the vertical position of a point in the string; thus, the second derivative $u_{tt}(x,t)$ is the acceleration.

If $u_{xx}(x,t) < 0$, i.e., if the string is concave down, the force (and hence the acceleration) point downward; thus $u_{tt}(x,t) < 0$; Figure 8.6a.

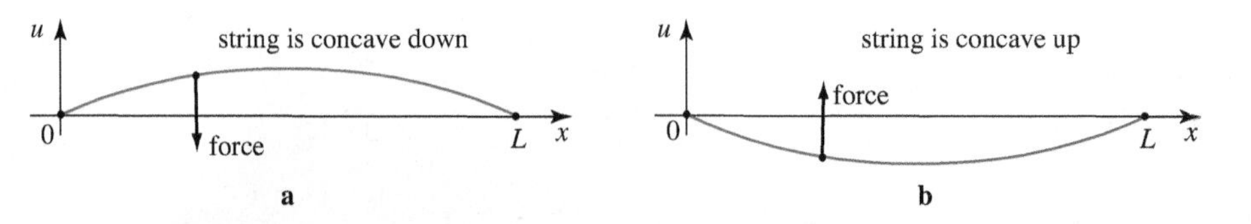

FIGURE 8.6

Relationship between u_{xx} and u_{tt}

If $u(x,t)$ is concave up, $u_{xx}(x,t) > 0$, and then the acceleration points upward, and $u_{tt}(x,t) > 0$; see Figure 8.6b. If $u(x,t)$ is flat (in a horizontal position, which means that $u_{xx}(x,t) = 0$), then it does not move and so the acceleration $u_{tt}(x,t)$ is zero.

As for a travelling wave, our analysis suggests a possible relationship between the two second partials. In this case, they are of the same sign, which is why we used a^2 in (8.2). There is more evidence of their possible relation—the larger the concavity, the stronger the force (i.e., the larger the acceleration) that is trying to bring the string back to its equilibrium position.

Example 8.3 Wave Equation

Show that $u(x,t) = \sin 2x \cos t$ satisfies the wave equation

$$u_{tt}(x,t) = \frac{1}{4}u_{xx}(x,t)$$

▶ This is a straightforward calculation. From $u_t(x,t) = -\sin 2x \sin t$ we get

$$u_{tt}(x,t) = -\sin 2x \cos t$$

and from $u_x(x,t) = 2\cos 2x \cos t$ we get

$$u_{xx}(x,t) = -4\sin 2x \cos t$$

It follows that

$$u_{tt}(x,t) = -\sin 2x \cos t = \frac{1}{4}(-4\sin 2x \cos t) = \frac{1}{4}u_{xx}(x,t)$$

▲

Example 8.4 Travelling Waves Satisfy the Wave Equation

The function $u(x,t) = f(x-st)$ describes a travelling wave of speed s whose shape is given by the graph of $y = f(x)$. Show that $u(x,t)$ satisfies the wave equation (8.2).

▶ The partial derivatives are computed by the chain rule:

$$u_x(x,t) = f'(x-st) \cdot 1 = f'(x-st)$$

and

$$u_{xx}(x,t) = f''(x-st) \cdot 1 = f''(x-st)$$

Using the chain rule again,

$$u_t(x,t) = f'(x-st)(-s) = -sf'(x-st)$$

and

$$u_{tt}(x,t) = -sf''(x-st)(-s) = s^2 f''(x-st)$$

We conclude that

$$u_{tt}(x,t) = s^2 f''(x-st) = s^2 u_{xx}(x,t)$$

Thus, $u(x,t)$ satisfies the wave equation (8.2) with $a^2 = s^2$.

▲

Diffusion Equation

Recall that we have already studied certain features of a diffusion process. As a model, we used the spread of a pollutant in the air. At the beginning of Section 1 we introduced the formula

$$c(x,t) = \frac{1}{\sqrt{4\pi Dt}} e^{-x^2/4Dt} \qquad (8.3)$$

for the concentration of the pollutant at time t and location x units away from the source of the pollution. The constant D is called the diffusion coefficient and identifies a specific diffusion process.

In Section 2 (Example 2.11) we investigated the features of the graph of $c(x,t)$ and interpreted various curves associated with it. In Section 4 (Example 4.5) we calculated and interpreted first-order partial derivatives of $c(x,t)$.

Diffusion is defined as the spread of particles in a medium from regions of higher concentration toward regions of lower concentration. Examples include heat diffusion, molecular diffusion, osmosis, facilitated diffusion (for instance across biological membranes), and so on. Equation (8.3) and the corresponding partial differential equation (which we discuss next) represent the simplest case, a building block on which other diffusion models are based. Even though our model is simple, it reveals essential and important properties of diffusion.

We continue using the spread of a pollutant as a model. Using our understanding of partial derivatives, we now investigate the relationship between the partial derivatives of $c(x,t)$.

Assume that at some fixed time t, the concentration $c(x,t)$ is represented by the graph in Figure 8.7a.

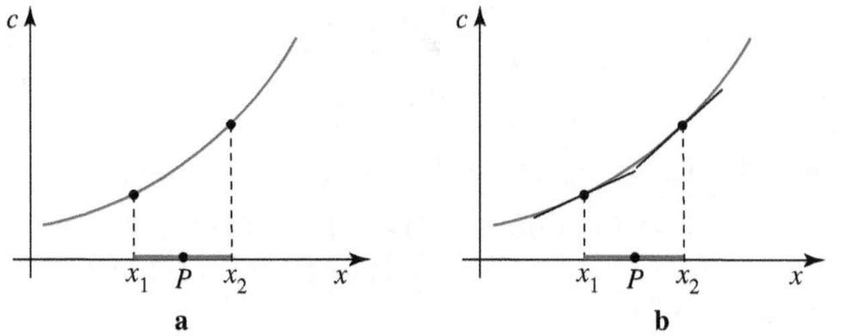

FIGURE 8.7

The graph of $c(x,t)$ for fixed t

Pick a location P and two neighbouring locations x_1 and x_2. From the graph we see that the concentration at x_2 is larger than the concentration at x_1. Since the pollutant spreads from regions of higher concentration toward regions of lower concentration, we conclude that the pollutant flows to the left, in the direction from x_2 to x_1. In other words, considering the segment $[x_1, x_2]$ on the x-axis, the pollutant flows into the segment at x_2 and leaves the segment at x_1.

What happens at P? Is the concentration increasing, or decreasing, or not changing?

According to *Fick's law* (from physics), the *rate* of the flow (the spread) of the pollutant is proportional to the partial derivative $c_x(x,t)$. This sounds logical: the larger *change* in concentration (i.e., the larger difference in concentration at two nearby points) will cause a stronger flow. Likewise, a smaller change in concentration will yield a weaker flow. (In the case of heat flow, this fact is called *Newton's law of cooling.*)

Looking at Figure 8.7b we see that the slope $c_x(x,t)$ at x_2 is larger than the slope at x_1. Thus, the flow across x_2 is stronger than the flow across x_1. This means that more pollutant enters the segment $[x_1, x_2]$ than leaves it, causing the concentration within the segment (and thus at P) to increase:

$$c_t(P,t) > 0$$

Note that the graph of c is concave up at P, and so

$$c_{xx}(P,t) > 0$$

Now consider the graph of $c(x,t)$ given in Figure 8.8 (again, the picture shows a snapshot of the concentration at some moment in time t). We would like to figure out whether, at Q, the concentration is increasing or decreasing.

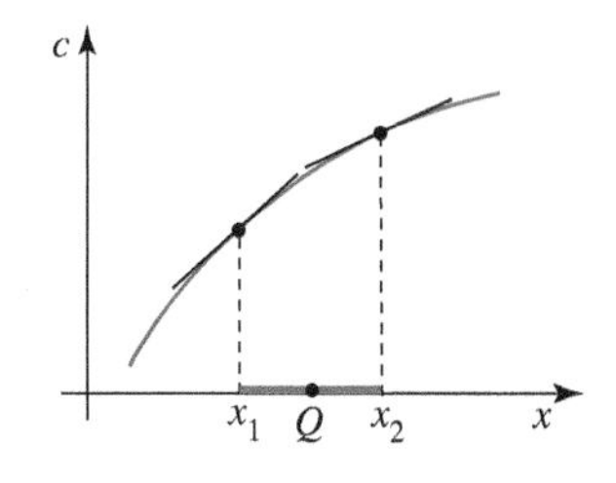

FIGURE 8.8

The graph of $c(x,t)$ for fixed t

Since $c(x_2,t) > c(x_1,t)$, the pollutant flows from the right to the left, as in the previous case. But this time

$$c_x(x_2,t) < c_x(x_1,t)$$

i.e., the slope at x_2 is smaller than the slope at x_1. Consequently, the flow at x_2 (where the pollutant enters $[x_1,x_2]$) is weaker than at at x_1 (where the pollutant leaves $[x_1,x_2]$), and so the concentration of the pollutant within the segment (and at Q) is decreasing:

$$c_t(Q,t) < 0$$

The graph of c is concave down at Q, and so

$$c_{xx}(Q,t) < 0$$

Consider one more case: $c(x,t)$, as a function of x (keep in mind that t is fixed), is linear and increasing; see Figure 8.9.

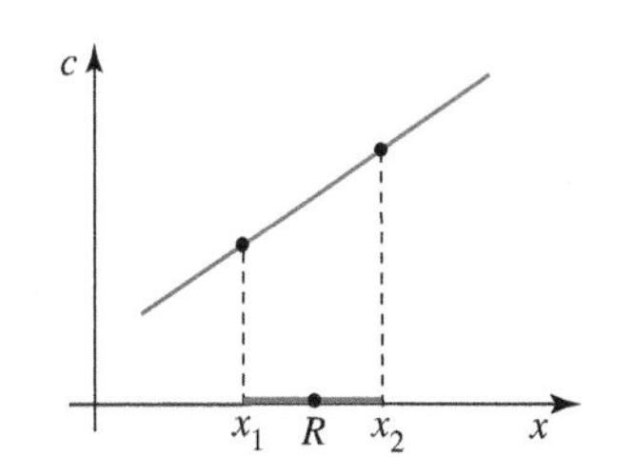

FIGURE 8.9

Linearly increasing concentration

Again, $c(x_2,t) > c(x_1,t)$ and so the pollutant flows from the right to the left. This time, the slopes at x_1 and at x_2 are equal, and so the flow into $[x_1,x_2]$ is identical to the flow out of $[x_1,x_2]$. Thus, there is no change in concentration at the location R, so

$$c_t(R,t) = 0$$

The line has zero concavity

$$c_{xx}(R,t) = 0$$

Looking at the three cases we have studied, we see that either c_t and c_{xx} have the same sign, or they are both zero. It turns out that the two are proportional:

$$c_t(x,t) = \sigma c_{xx}(x,t) \tag{8.4}$$

for a positive constant σ. (The proof of this fact is beyond what we can do in this book.)

The partial differential equation (8.4) is called the *diffusion equation* or, in some situations, the *heat equation*. The constant of proportionality σ depends on the diffusion process that we are studying.

Example 8.5 Example of a Function Satisfying the Diffusion Equation

Show that the function $c(x,t) = a + e^{-bt}\sin(mx)$, where a, b, and m are constants and $m \neq 0$, satisfies the diffusion equation (8.4). What is the value of σ?

▶ We compute

$$c_t(x,t) = e^{-bt}(-b)\sin(mx) = -be^{-bt}\sin(mx)$$
$$c_x(x,t) = e^{-bt}\cos(mx)(m) = me^{-bt}\cos(mx)$$

and

$$c_{xx}(x,t) = me^{-bt}(-\sin(mx))(m) = -m^2e^{-bt}\sin(mx)$$

Thus

$$c_t(x,t) = -be^{-bt}\sin(mx) = \frac{b}{m^2}\left[-m^2e^{-bt}\sin(mx)\right] = \frac{b}{m^2}c_{xx}(x,t)$$

(we divided and multiplied by m^2 so that we could recognize the expression for c_{xx}). So, $c(x,t)$ satisfies equation (8.4) with $\sigma = b/m^2$. ▲

Example 8.6 Diffusion Equation

Show that the concentration (8.3)

$$c(x,t) = \frac{1}{\sqrt{4\pi Dt}} e^{-x^2/4Dt}$$

satisfies the diffusion equation (8.4).

▶ This is not difficult, just a bit messy. First, rewrite $c(x,t)$ to separate the constants from the variables

$$c(x,t) = \frac{1}{\sqrt{4\pi D}} t^{-1/2} e^{-\frac{1}{4D}x^2t^{-1}}$$

Using the product rule and the chain rule

$$\begin{aligned} c_t(x,t) &= \frac{1}{\sqrt{4\pi D}}\left(-\frac{1}{2}\right) t^{-3/2} e^{-\frac{1}{4D}x^2t^{-1}} + \frac{1}{\sqrt{4\pi D}} t^{-1/2} e^{-\frac{1}{4D}x^2t^{-1}} \left(-\frac{1}{4D}\right) x^2 \left(-t^{-2}\right) \\ &= \left(-\frac{1}{2}\right) \frac{1}{\sqrt{4\pi D}} t^{-3/2} e^{-\frac{1}{4D}x^2t^{-1}} + \frac{1}{4D} \frac{1}{\sqrt{4\pi D}} x^2 t^{-5/2} e^{-\frac{1}{4D}x^2t^{-1}} \\ &= \left(-\frac{1}{2}\right) \frac{1}{\sqrt{4\pi D}} t^{-3/2} e^{-\frac{1}{4D}x^2t^{-1}} \left(1 - \frac{1}{2D}x^2t^{-1}\right) \end{aligned}$$

Next, we calculate the partials with respect to x:

$$\begin{aligned} c_x(x,t) &\frac{1}{\sqrt{4\pi D}} t^{-1/2} e^{-\frac{1}{4D}x^2t^{-1}} \left(-\frac{1}{4D}\right) 2xt^{-1} \\ &= \left(-\frac{1}{2D}\right) \frac{1}{\sqrt{4\pi D}} t^{-3/2}\, x\, e^{-\frac{1}{4D}x^2t^{-1}} \end{aligned}$$

and

$$\begin{aligned} c_{xx}(x,t) &= \left(-\frac{1}{2D}\right) \frac{1}{\sqrt{4\pi D}} t^{-3/2} \left(e^{-\frac{1}{4D}x^2t^{-1}} + xe^{-\frac{1}{4D}x^2t^{-1}} \left(-\frac{1}{4D}\right) 2xt^{-1}\right) \\ &= \left(-\frac{1}{2D}\right) \frac{1}{\sqrt{4\pi D}} t^{-3/2} e^{-\frac{1}{4D}x^2t^{-1}} \left(1 - \frac{1}{2D}x^2t^{-1}\right) \\ &= \frac{1}{D} c_t(x,t) \end{aligned}$$

Thus, $c_t(x,t) = Dc_{xx}(x,t)$, and we are done. ▲

The partial differential equation (8.4) and the solution $c(x,t)$ that we verified in Example 8.6 represent the one-dimensional diffusion equation, since there is only one space variable (x). It can be generalized to two dimensions, in which case we use the function $c(x,y,t)$ to describe the concentration at the location (x,y) at time t; the corresponding partial differential equation is

$$c_t(x,y,t) = \sigma\left(c_{xx}(x,y,t) + c_{yy}(x,y,t)\right)$$

or simply $c_t = \sigma(c_{xx} + c_{yy})$; σ is a positive constant. This equation can be further expanded to include diffusion in three dimensions.

The related partial differential equation

$$c_{xx}(x,y) + c_{yy}(x,y) = 0 \qquad (8.5)$$

is called *Laplace's equation* and its solutions are refered to as *harmonic functions.* Laplace's equation is used to study numerous applications such as fluid flow, flow of material within a cell, or growth of certain types of cancer.

For a moment, we will go back to the solution

$$c(x,t) = \frac{1}{\sqrt{4\pi Dt}} e^{-x^2/4Dt}$$

of the diffusion equation. For fixed t, the shape of the graph of $c(x,t)$ is the *Gaussian distribution function* or *Gaussian density function*; see Figure 8.10.

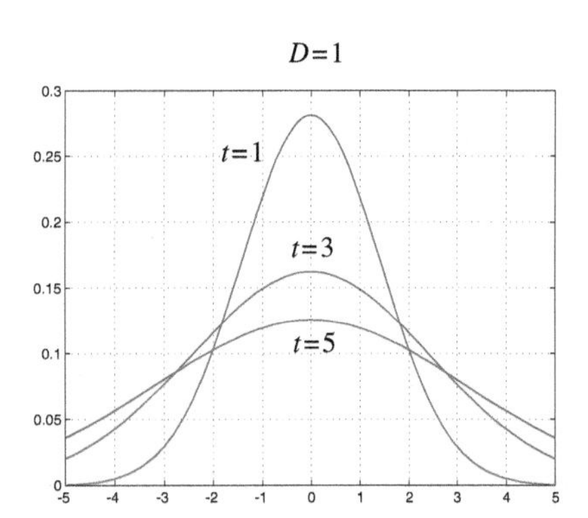

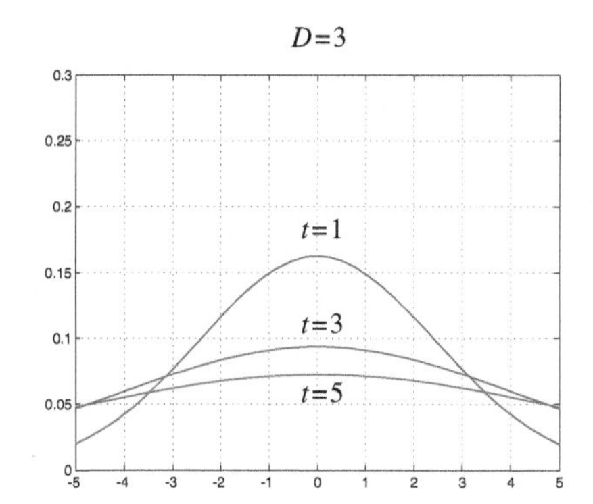

FIGURE 8.10

Comparing the speed of diffusion

The diffusion coefficient D determines the speed of the diffusion process. In Figure 8.10 we have taken snapshots of the diffusion process (graph of $c(x,t)$) at times $t=1$, $t=3$, and $t=5$ in cases where $D=1$ and $D=3$. The graphs indicate that a larger value for D implies faster diffusion (spread).

Certain types of diffusion are very slow. For instance, an oxygen molecule in the blood takes about 8 minutes to move a distance of 1 mm (solely by diffusion). That's why the human body uses other mechanisms to move material (not just oxygen) around.

Summary **Partial differential equations** help us model all kinds of phenomena. Using the first partials, we describe a **travelling wave,** a type of horizontal movement of water (or other particles) where the shape of the wave is preserved. A **one-dimensional wave equation,** explained on a vibrating string, can be applied to investigate periodic motions such as sound waves, the flow of blood, or movements of air masses in the atmosphere. The **diffusion equation** models a specific type of particle spread within a medium.

8 Exercises

1. Explain why $u(x,t) = (x-3t)^2 - 1$ represents a travelling wave. Identify its speed.

2. We know that if f is a differentiable function of one variable, then $u(x,t) = f(x-st)$ is a travelling wave of speed s. Is $v(x,t) = f(x+st)$ a travelling wave? If so, what is its speed?

3–8 ▪ Verify that the following functions are harmonic, i.e., that they satisfy Laplace's equation $f_{xx} + f_{yy} = 0$.

3. $f(x,y) = x^4 - 6x^2y^2 + y^4$

4. $f(x,y) = 5x - y + 12$

5. $f(x,y) = e^x \sin y$

6. $f(x,y) = \ln(x^2+y^2)$

7. $f(x,y) = \arctan(y/x)$

8. $f(x,y) = x^3 - 3xy^2$

9. Show that $c(x,t) = e^{Ax+Bt}$ (A and B are constants) satisfies the diffusion equation (8.4). What is the value of σ?

10. Show that $c(x,t) = t^{-1/2}e^{-x^2/t}$ satisfies the diffusion equation (8.4). What is the value of σ?

11. Show that a travelling wave (i.e., a twice differentiable function $u(x,t)$ such that $u_t(x,t) = -su_x(x,t)$) satisfies the wave equation $u_{tt}(x,t) = a^2u_{xx}(x,t)$. What is the value of a^2?

12. Prove that the function $c(x,t) = e^{-k^2 t} \sin kx$ satisfies the heat/diffusion equation (8.4). What is the value of σ?

13. Verify that the function $u(x,t) = \sin 2x \sin 3t$ satisfies the wave equation (8.2). What is the value of a^2?

14. Verify that the function $u(x,y,t) = \sin 2x \sin 2y \sin 3t$ satisfies the two-dimensional wave equation $u_{tt} = a^2(u_{xx} + u_{yy})$. What is the value of a^2?

15. Verify that the function $u(x,t) = e^{x-2t} + (x+2t)^2$ satisfies the wave equation (8.2). What is the value of a^2?

16. Take any two functions f and g of one variable. Assuming that the second derivatives f'' and g'' exist, prove that $u(x,t) = f(x-kt) + g(x+kt)$ (where k is a constant) satisfies the wave equation (8.2).

17. Consider the motion of a string given by $u(x,t) = \sin x \sin 2t$, $0 \leq x \leq \pi$.

    (a) Find $u_{xx}(\pi/4, 0.5)$ and interpret your answer in terms of the motion of the string.

    (b) Find $u_t(\pi/4, 0.5)$ and $u_{tt}(\pi/4, 0.5)$ and interpret your answer.

18. The temperature at time t at a point x in a metal rod of length 1 placed along the x-axis with one end at the origin is given by $T(x,t) = 1 + e^{-t} \sin \pi x$.

    (a) What is the initial temperature at the ends of the rod? Initially, what is the warmest point on the rod?

    (b) Show that $T(x,t)$ satisfies the heat equation $T_t = T_{xx}$ (with $\sigma = 1/\pi^2$).

    (c) Sketch the graph of the temperature $T(x,t)$ at times $t = 0$, $t = 1$, and $t = 2$ in the same coordinate system. What happens as $t \to \infty$?

9 Directional Derivative and Gradient

The partial derivatives f_x and f_y of a function f tell us how f changes in two special directions: parallel to the x-axis and parallel to the y-axis. Now we learn how to calculate the rate of change of f in an arbitrary direction by using the **directional derivative.** By investigating the properties of the directional derivative, we discover the **gradient vector.** Built of partial derivatives, the gradient vector carries essential information about the way a function changes.

We review relevant facts about vectors in the Appendix at the end of this section.

Directional Derivative

We start with an example. Assume that the function $T(x, y)$ represents the temperature in degrees Celsius at a location (x, y), where x (eastern direction) and y (northern direction) are measured in kilometres. The level curves of T are drawn in Figure 9.1.

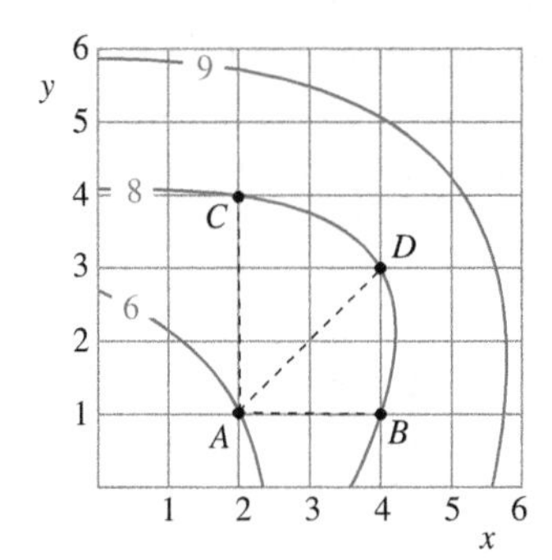

FIGURE 9.1

Level curves of the temperature function $T(x, y)$

At A, the temperature is 6°C. Walking in the horizontal direction away from A for 2 km, we arrive at the point B, where we meet the level curve of value 8°C. The average rate of change in the horizontal direction from A to B is

$$\frac{T(B) - T(A)}{2} = \frac{8 - 6}{2} = 1\,^{\circ}\mathrm{C/km}$$

This number is also an estimate for the partial derivative, i.e., $\partial T/\partial x(A) \approx 1\,^{\circ}\mathrm{C/km}$. The average rate of change from A to C is

$$\frac{T(C) - T(A)}{3} = \frac{8 - 6}{3} = \frac{2}{3} \approx 0.667\,^{\circ}\mathrm{C/km}$$

and $\partial T/\partial y(A) \approx 2/3\,^{\circ}\mathrm{C/km}$. We conclude that it gets warmer more quickly if we walk west than if we walk east.

In the same way we can compute the average rate of change in the temperature as we walk northeast from A. After walking for $\sqrt{2^2 + 2^2} = \sqrt{8} = 2\sqrt{2}$ km, we meet the level curve 8°C at D. Thus, the average rate of change at A in the northwest direction is

$$\frac{T(D) - T(A)}{2\sqrt{2}} = \frac{8 - 6}{2\sqrt{2}} = \frac{1}{\sqrt{2}} \approx 0.707\,^{\circ}\mathrm{C/km}$$

Which derivative is this an estimate for?

It is an estimate for the *directional derivative,* which we now define. The directional derivative will allow us to calculate the rate of change of a function in an arbitrary direction.

To describe a direction, we use a unit vector $\mathbf{u}$; "unit" means that its length is 1, i.e., $\|\mathbf{u}\| = 1$. (See the Appendix at the end of this section, where we review basic facts about vectors.) The direction of the positive x-axis is given by the vector $\mathbf{u} = \mathbf{i}$, and the direction of the positive y-axis is given by $\mathbf{u} = \mathbf{j}$. Given two points $P(p_1, p_2)$ and $Q(q_1, q_2)$, the vector $\mathbf{v}$ from P to Q is given by $\mathbf{v} = (q_1 - p_1)\mathbf{i} + (q_2 - p_2)\mathbf{j}$. Its length

$$\|\mathbf{v}\| = \sqrt{(q_1 - p_1)^2 + (q_2 - p_2)^2}$$

is equal to the distance from P to Q. If $P \neq Q$, then $\mathbf{v}$ is not a zero vector, and hence it has a direction. As well, $\|\mathbf{v}\| > 0$. The corresponding unit vector (i.e., the vector of length 1 and of the same direction as $\mathbf{v}$) is given by

$$\mathbf{u} = \frac{\mathbf{v}}{\|\mathbf{v}\|} = \frac{(q_1 - p_1)\mathbf{i} + (q_2 - p_2)\mathbf{j}}{\sqrt{(q_1 - p_1)^2 + (q_2 - p_2)^2}}$$

In this way, we can find a unit vector in any direction.

As an illustration, we find a unit vector in the direction from $A(2,1)$ to $D(4,3)$ in Figure 9.1. The vector from A to D is $\mathbf{v} = (4-2)\mathbf{i} + (3-1)\mathbf{j} = 2\mathbf{i} + 2\mathbf{j}$. Its norm is $\|\mathbf{v}\| = \sqrt{2^2 + 2^2} = \sqrt{8} = 2\sqrt{2}$ and the unit vector in the direction of $\mathbf{v}$ is

$$\mathbf{u} = \frac{\mathbf{v}}{\|\mathbf{v}\|} = \frac{2\mathbf{i} + 2\mathbf{j}}{2\sqrt{2}} = \frac{1}{\sqrt{2}}\mathbf{i} + \frac{1}{\sqrt{2}}\mathbf{j}$$

Take a function $f(x, y)$ and a point (a, b) in its domain. We now calculate the rate of change of f in the direction given by the unit vector $\mathbf{u} = u_1\mathbf{i} + u_2\mathbf{j}$.

As for any rate of change we have calculated, we need a nearby point. For $h > 0$, the vector $h\mathbf{u} = hu_1\mathbf{i} + hu_2\mathbf{j}$ points in the same direction as $\mathbf{u}$, and its length is $\|h\mathbf{u}\| = |h|\,\|\mathbf{u}\| = h$ (since $\|\mathbf{u}\| = 1$). Thus, the point P with coordinates $(a + hu_1, b + hu_2)$ is h units away from A; see Figure 9.2.

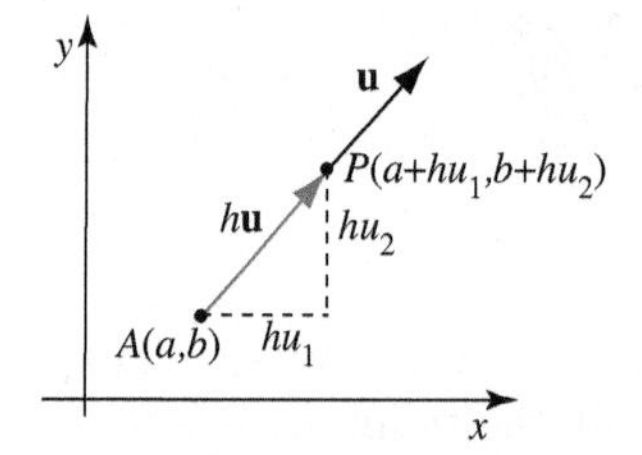

FIGURE 9.2
Investigating rate of change

We now define the average rate of change from A to P in the usual way:

$$\frac{\text{change in } f \text{ from } A \text{ to } P}{\text{distance from } A \text{ to } P} = \frac{f(a + hu_1, b + hu_2) - f(a, b)}{h}$$

Taking the limit gives the instantaneous rate of change.

Definition 16 Directional Derivative

The *directional derivative* of a function $f(x, y)$ at a point (a, b) in the direction of a unit vector $\mathbf{u} = u_1\mathbf{i} + u_2\mathbf{j}$ is given by

$$D_{\mathbf{u}}f(a, b) = \lim_{h \to 0} \frac{f(a + hu_1, b + hu_2) - f(a, b)}{h}$$

provided that the limit exists. ▲

The notation $D_{\mathbf{u}}f(a, b)$ for the directional derivative contains all the needed information: the function f, the point (a, b), and the unit direction $\mathbf{u}$. Keep in mind that the definition requires that we use a unit vector. Even though we might say "in the direction of the vector $2\mathbf{i} + 2\mathbf{j}$," in all calculations we need to convert the given direction into a unit vector.

The directional derivative $D_{\mathbf{u}}f(a, b)$ represents the rate of change of f in the given direction, in the same sense as all derivatives do.

For instance, if $D_{\mathbf{u}}f(a, b) > 0$, then f is increasing at (a, b) in the direction of $\mathbf{u}$; if $D_{\mathbf{u}}f(a, b) < 0$, then f is decreasing at (a, b) in the direction of $\mathbf{u}$, and so on.

Definition 16 tells us that the estimate of 0.707 °C/km that we arrived at in the introductory example is an estimate for $D_{\mathbf{u}}T(3, 1)$, where $\mathbf{u} = 1/\sqrt{2}\mathbf{i} + 1/\sqrt{2}\mathbf{j}$.

If $\mathbf{u} = \mathbf{i} = 1 \cdot \mathbf{i} + 0 \cdot \mathbf{j}$, then

$$\begin{aligned} D_{\mathbf{i}}f(a, b) &= \lim_{h \to 0} \frac{f(a + h(1), b + h(0)) - f(a, b)}{h} \\ &= \lim_{h \to 0} \frac{f(a + h, b) - f(a, b)}{h} \\ &= \frac{\partial f}{\partial x}(a, b) \end{aligned}$$

Likewise, $D_{\mathbf{j}}f(a, b) = \partial f/\partial y(a, b)$. Thus, the directional derivative of a function generalizes its partial derivatives.

Example 9.1 Calculating the Directional Derivative

Find the rate of change of $f(x, y) = y^2 - x$ at $(1, 2)$ in the direction of the vector $\mathbf{v} = 3\mathbf{i} + 4\mathbf{j}$.

▶ The vector $\mathbf{v}$ is not a unit vector, since $\|\mathbf{v}\| = \sqrt{3^2 + 4^2} = 5$. The unit vector in the direction of $\mathbf{v}$ is

$$\mathbf{u} = \frac{\mathbf{v}}{\|\mathbf{v}\|} = \frac{3}{5}\mathbf{i} + \frac{4}{5}\mathbf{j}$$

We are asked to calculate $D_{\mathbf{u}}f(1, 2)$. Using Definition 16, we write

$$\begin{aligned} D_{\mathbf{u}}f(1, 2) &= \lim_{h \to 0} \frac{f(1 + h(3/5), 2 + h(4/5)) - f(1, 2)}{h} \\ &= \lim_{h \to 0} \frac{1}{h}\left(\left[\left(2 + \frac{4h}{5}\right)^2 - \left(1 + \frac{3h}{5}\right)\right] - [2^2 - 1]\right) \\ &= \lim_{h \to 0} \frac{1}{h}\left(4 + \frac{16h}{5} + \frac{16h^2}{25} - 1 - \frac{3h}{5} - 3\right) \\ &= \lim_{h \to 0} \frac{1}{h}\left(\frac{13h}{5} + \frac{16h^2}{25}\right) \\ &= \lim_{h \to 0}\left(\frac{13}{5} + \frac{16h}{25}\right) = \frac{13}{5} \end{aligned}$$

Thus, at $(1, 2)$, in the direction of the given vector, the function increases at a rate of $13/5$ units per unit increase in distance. ▲

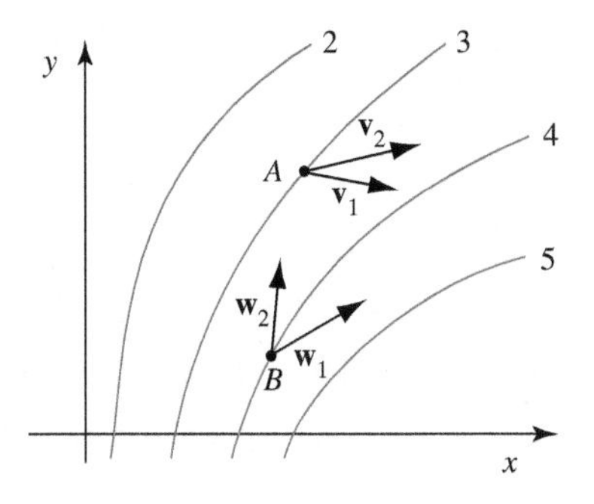

FIGURE 9.3

Level curves and vectors of Example 9.2

Example 9.2 Geometric Reasoning about Directional Derivatives

Figure 9.3 shows level curves of a function $f(x, y)$. All vectors are assumed to be of unit length. Which of the two directional derivatives $D_{\mathbf{v}_1} f(A)$ and $D_{\mathbf{v}_2} f(A)$ is larger? Determine whether $D_{\mathbf{w}_1} f(B)$ and $D_{\mathbf{w}_2} f(B)$ are positive or negative.

▶ The value of f at A is 3. Both $\mathbf{v}_1$ and $\mathbf{v}_2$ point toward the contour curve of value 4, so f is increasing in both directions. Walking in the direction of $\mathbf{v}_1$, we reach the level curve 4 sooner than walking (at the same speed) in the direction of $\mathbf{v}_2$. In other words, the rate of change in the direction of $\mathbf{v}_1$ is larger, since the change in values of f from 3 to 4 happens over a smaller distance. Thus, $D_{\mathbf{v}_1} f(A) > D_{\mathbf{v}_2} f(A)$.

At B, the vector $\mathbf{w}_1$ points toward increasing values of f, and thus $D_{\mathbf{w}_1} f(B) > 0$. The vector $\mathbf{w}_2$ points toward the level curve of value 3, indicating that f is decreasing; thus $D_{\mathbf{w}_2} f(B) < 0$. ▲

Example 9.3 Directional Derivative along a Level Curve

At a point A, a unit vector $\mathbf{u}$ is tangent to the level curve of a differentiable function f; see Figure 9.4. What is the value of $D_{\mathbf{u}} f(A)$?

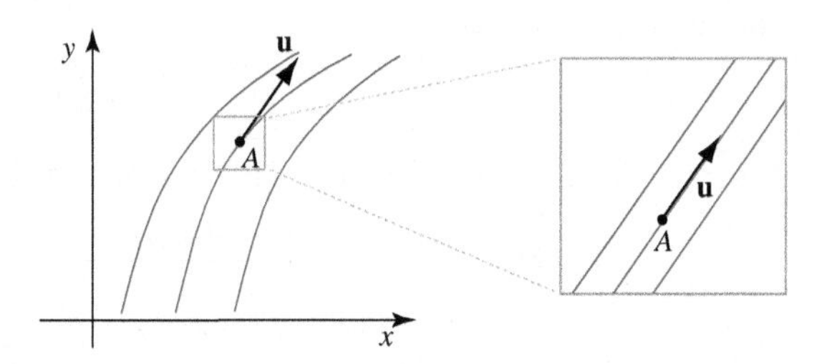

FIGURE 9.4
Level curves of f and magnification near A

▶ Recall that, locally, the level curves of a differentiable function look like the level curves of its linear approximation: an equally spaced set of parallel lines. So, zooming in on the contour curves of f near A, we obtain the diagram shown in Figure 9.4. We see that moving in the direction of the tangent will not produce a change in f, and so $D_{\mathbf{u}} f(A) = 0$. ▲

Calculating a derivative from the limit definition is, quite often, complicated and difficult. Our aim now is to find an easier, more convenient way to calculate directional derivatives (at least in some cases).

Assume that f is differentiable. Definition 16 states that

$$D_{\mathbf{u}} f(a, b) = \lim_{h \to 0} \frac{f(a + hu_1, b + hu_2) - f(a, b)}{h} \tag{9.1}$$

Recall the linear approximation of f: for (x, y) near (a, b),

$$f(x, y) \approx f(a, b) + f_x(a, b)(x - a) + f_y(a, b)(y - b) \tag{9.2}$$

(see Definition 12 in Section 5). In our case, the nearby point is $(x, y) = (a + hu_1, b + hu_2)$; i.e., $x = a + hu_1$ and $y = b + hu_2$. Thus,

$$\begin{aligned} f(a + hu_1, b + hu_2) &\approx f(a, b) + f_x(a, b)((a + hu_1) - a) + f_y(a, b)((b + hu_2) - b) \\ &= f(a, b) + f_x(a, b)hu_1 + f_y(a, b)hu_2 \end{aligned}$$

and

$$f(a + hu_1, b + hu_2) - f(a, b) \approx f_x(a, b)hu_1 + f_y(a, b)hu_2$$

The difference quotient in (9.1) is approximated by

$$\frac{f(a+hu_1,b+hu_2)-f(a,b)}{h} \approx \frac{f_x(a,b)hu_1+f_y(a,b)hu_2}{h}$$
$$= f_x(a,b)u_1+f_y(a,b)u_2$$

Taking the limit as $h \to 0$, we obtain

$$D_{\mathbf{u}}f(a,b) = \lim_{h\to 0}\frac{f(a+hu_1,b+hu_2)-f(a,b)}{h}$$
$$= f_x(a,b)u_1+f_y(a,b)u_2$$

Thus, we have finished the proof of the following theorem.

Theorem 14 Directional Derivative

Assume that f is a differentiable function and $\mathbf{u} = u_1\mathbf{i} + u_2\mathbf{j}$ a unit vector. The directional derivative of f at a point (a,b) in the direction $\mathbf{u}$ is given by

$$D_{\mathbf{u}}f(a,b) = f_x(a,b)u_1 + f_y(a,b)u_2$$

Thus, the directional derivative of f in a general direction can be expressed as a combination of the two partial derivatives f_x and f_y.

Example 9.4 Calculating a Directional Derivative

Going back to Example 9.1, we recalculate $D_{\mathbf{u}}f(1,2)$ for $f(x,y) = y^2 - x$ in the direction of the vector $\mathbf{u} = 3/5\mathbf{i} + 4/5\mathbf{j}$.

▶ The partial derivatives of f are $f_x = -1$ and $f_y = 2y$. Thus, $f_x(1,2) = -1$, $f_y(1,2) = 4$, and

$$D_{\mathbf{u}}f(1,2) = f_x(1,2)u_1 + f_y(1,2)u_2 = -1\cdot\frac{3}{5} + 4\cdot\frac{4}{5} = \frac{13}{5}$$

Example 9.5 Calculating and Interpreting Directional Derivatives

The air pressure within a certain region in a plane is given by the formula $P(x,y) = 3x^2y - y^4 + 2$. At the point $(3,1)$, in which of the three directions, $\mathbf{i}$, $\mathbf{j}$, or $3\mathbf{i}+2\mathbf{j}$, is the pressure increasing the most quickly?

▶ The partial derivatives of P are $P_x = 6xy$ and $P_y = 3x^2 - 4y^3$; evaluated at the given point, $P_x(3,1) = 18$ (this is the derivative in the direction of $\mathbf{i}$) and $P_y(3,1) = 23$ (this is the derivative in the direction of $\mathbf{j}$).

The unit vector in the direction of $3\mathbf{i}+2\mathbf{j}$ is

$$\mathbf{u} = \frac{3\mathbf{i}+2\mathbf{j}}{\|3\mathbf{i}+2\mathbf{j}\|} = \frac{3\mathbf{i}+2\mathbf{j}}{\sqrt{13}} = \frac{3}{\sqrt{13}}\mathbf{i} + \frac{2}{\sqrt{13}}\mathbf{j}$$

and so

$$D_{\mathbf{u}}P(3,1) = P_x(3,1)u_1 + P_y(3,1)u_2$$
$$= 18\cdot\frac{3}{\sqrt{13}} + 23\cdot\frac{2}{\sqrt{13}}$$
$$= \frac{100}{\sqrt{13}} \approx 27.735$$

Thus, P changes most quickly (at a rate of 27.735 units of pressure per unit of distance) in the direction of $3\mathbf{i}+2\mathbf{j}$.

The Gradient Vector

We can think of the partial derivatives of a differentiable function as components of a vector. This is not just a convenience but a very useful concept.

Definition 17 The Gradient Vector

Assume that f is a differentiable function. The *gradient* of f at (x, y) is the vector $\nabla f(x, y)$ defined by

$$\nabla f(x, y) = f_x(x, y)\,\mathbf{i} + f_y(x, y)\,\mathbf{j}$$

Sometimes, the notation grad$f$ is used instead of ∇f.

For instance, if $f(x, y) = x^2y + y^2 - 3$, then

$$\nabla f(x, y) = 2xy\mathbf{i} + (x^2 + 2y)\mathbf{j}$$

As a matter of fact, $\nabla f(x, y)$ is a vector function, since its components are functions; it is called a *vector field* (we defined and graphed vector fields near the end of Section 2). But when we specify a point, we get a true vector. For instance, $\nabla f(1, 1) = 2\mathbf{i} + 3\mathbf{j}$, $\nabla f(0, 1) = 2\mathbf{j}$, $\nabla f(-1, -1) = 2\mathbf{i} - \mathbf{j}$, $\nabla f(-2, 0) = 4\mathbf{j}$, and so on. In Figure 9.5 we drew the four gradient vectors how they are usually drawn: with the tail located at the point where the vector is calculated.

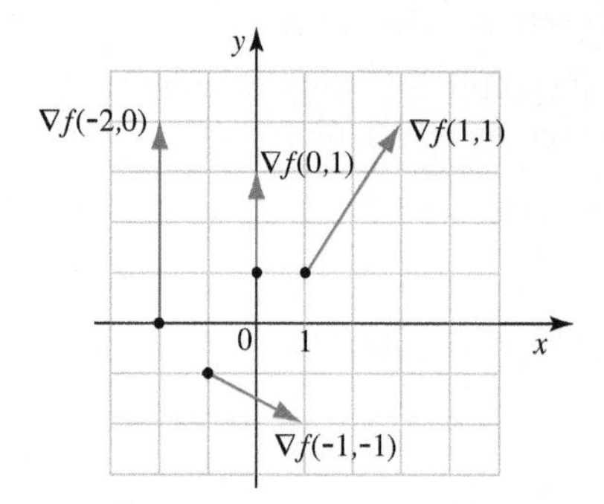

FIGURE 9.5

Gradient vectors at several points in the plane

Recall that the dot product of two vectors $\mathbf{v} = v_1\mathbf{i} + v_2\mathbf{j}$ and $\mathbf{w} = w_1\mathbf{i} + w_2\mathbf{j}$ is the real number

$$\mathbf{v} \cdot \mathbf{w} = v_1w_1 + v_2w_2$$

With this fact in mind, we recognize the formula in Theorem 14 as the dot product of the gradient vector $\nabla f(a, b) = f_x(a, b)\mathbf{i} + f_y(a, b)\mathbf{j}$ and the unit vector $\mathbf{u} = u_1\mathbf{i} + u_2\mathbf{j}$:

$$\begin{aligned} D_{\mathbf{u}}f(a, b) &= f_x(a, b)u_1 + f_y(a, b)u_2 \\ &= \big(f_x(a, b)\mathbf{i} + f_y(a, b)\mathbf{j}\big) \cdot \big(u_1\mathbf{i} + u_2\mathbf{j}\big) \\ &= \nabla f(a, b) \cdot \mathbf{u} \end{aligned} \tag{9.3}$$

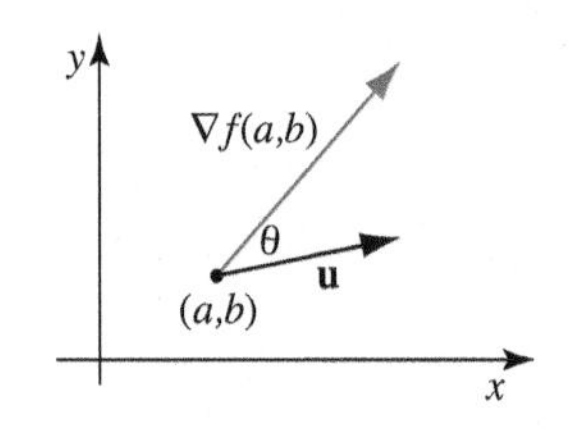

FIGURE 9.6

Directions in the plane relative to $\nabla f(a, b)$

Take a differential function $f(x, y)$. With a point (a, b) in its domain we associate the vector $\nabla f(a, b)$; see Figure 9.6. In this way, if $\nabla f(a, b)$ is not a zero vector, we obtain a canonical (i.e., default) direction that we use as a reference. Any other direction in the plane can be identified by specifying the angle with respect to $\nabla f(a, b)$.

Pick a unit vector $\mathbf{u}$ that makes an angle θ with respect to $\nabla f(a, b)$. (Recall that the angle between two vectors is defined to be the smaller of the two angles formed by their directions; thus, $0 \leq \theta \leq \pi$.) Using the geometric formula for the dot product (see (9.5) in the Appendix at the end of this section), we write

$$D_{\mathbf{u}}f(a,b) = \nabla f(a,b) \cdot \mathbf{u} = \|\nabla f(a,b)\| \, \|\mathbf{u}\| \cos\theta = \|\nabla f(a,b)\| \cos\theta \qquad (9.4)$$

since $\|\mathbf{u}\| = 1$.

The formula $D_{\mathbf{u}}f(a,b) = \|\nabla f(a,b)\| \cos\theta$ gives an easy way to calculate the directional derivative: we multiply the length of the gradient vector at the given point by the cosine of the angle that the desired direction makes with the gradient vector.

Example 9.6 Gradient and Directional Derivatives

Let $f(x,y) = 2xe^{y-1} + x^2y$. The gradient of f is

$$\nabla f(x,y) = \left(2e^{y-1} + 2xy\right)\mathbf{i} + \left(2xe^{y-1} + x^2\right)\mathbf{j}$$

Pick a point, say, $(1,1)$. Then $\nabla f(1,1) = 4\mathbf{i} + 3\mathbf{j}$ and $\|\nabla f(1,1)\| = \sqrt{4^2 + 3^2} = 5$. Thus

$$D_{\mathbf{u}}f(1,1) = \|\nabla f(1,1)\| \cos\theta = 5\cos\theta$$

where θ is the angle measured from $\nabla f(1,1)$. For instance, the derivative in the direction $\pi/8$ radians from the gradient vector $\nabla f(1,1)$ is

$$D_{\theta=\pi/8}f(1,1) = 5\cos(\pi/8) \approx 4.619$$

The derivative in the direction of the gradient is

$$D_{\theta=0}f(1,1) = 5\cos 0 = 5$$

and so on. In Figure 9.7 we show equally spaced directions ($\pi/8$ radians apart) and the corresponding rates of change of f. ▲

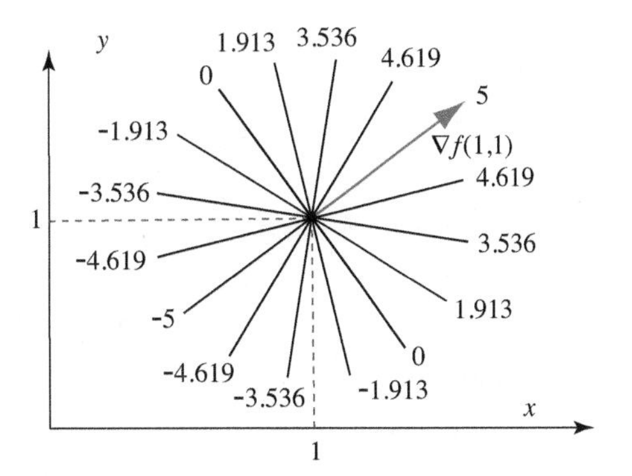

FIGURE 9.7

Rates of change of f in various directions

Obviously, the rates differ, depending on the direction. So in which direction is the rate of change the largest? The smallest?

Figure 9.7 suggests that the largest rate of change occurs in the direction of the gradient; in the opposite direction, the function decreases the most.

Let's see why this is true. In formula (9.4)

$$D_{\mathbf{u}}f(a,b) = \|\nabla f(a,b)\| \cos\theta$$

the quantity $\|\nabla f(a,b)\|$ is fixed; the angle θ is the only variable. Since $\cos\theta \leq 1$, the largest value of $D_{\mathbf{u}}f(a,b)$ occurs when $\cos\theta = 1$, i.e., when $\theta = 0$. In that case,

$$D_{\theta=0}f(a,b) = \|\nabla f(a,b)\| \cos 0 = \|\nabla f(a,b)\|$$

The term $\cos\theta$ reaches its smallest value of -1 when $\theta = \pi$; in that case,

$$D_{\theta=\pi}f(a,b) = \|\nabla f(a,b)\| \cos\pi = -\|\nabla f(a,b)\|$$

When $\theta = \pi/2$, then

$$D_{\theta=\pi/2}f(a,b) = \|\nabla f(a,b)\| \cos(\pi/2) = 0$$

We have just discovered an important result.

Theorem 15 Extreme Values of the Directional Derivative

Assume that f is a differentiable function, and (a, b) is a point in its domain where $\nabla f(a,b) \neq \mathbf{0}$.

(a) The maximum rate of change of f at (a, b) is equal to $\|\nabla f(a,b)\|$ and occurs in the direction of the gradient $\nabla f(a,b)$.

(b) The minimum rate of change of f at (a, b) is equal to $-\|\nabla f(a,b)\|$ and occurs in the direction opposite to the gradient $\nabla f(a,b)$.

In the directions perpendicular to the gradient, the directional derivative is zero. Note that if $\nabla f(a,b) = \mathbf{0}$, then $D_{\mathbf{u}} f(a,b) = 0$ in all directions (in this case, the point (a, b) is called a *critical point* of f; in Section 10 we study critical points in detail).

In Example 11.4 in Section 11 we give an alternative proof of the theorem.

Example 9.7 Working with Directional Derivatives

Assume that the function $h(x,y) = 1200 - x^2 - 2y^2$ models the height of a hill (in metres) at a location (x, y). Thus, at the point $(10, 4)$, the height is $h(10,4) = 1200 - 10^2 - 2(4)^2 = 1068$ m. The coordinates x and y are measured in kilometres.

Identify the direction of the steepest ascent from the point $(10, 4)$. As well, identify all directions at $(10, 4)$ in which the slope of the hill is larger than 70% of the maximum slope.

From $\nabla h = -2x\mathbf{i} - 4y\mathbf{j}$, we get $\nabla h(10,4) = -20\mathbf{i} - 16\mathbf{j}$ and

$$\|\nabla h(10,4)\| = \sqrt{20^2 + 16^2} = \sqrt{656} = 4\sqrt{41}$$

The direction of the steepest ascent (maximim rate of change of h) is given by the gradient $\nabla h(10,4) = -20\mathbf{i} - 16\mathbf{j}$; expressed as a unit vector,

$$\mathbf{u} = \frac{-20\mathbf{i} - 16\mathbf{j}}{4\sqrt{41}} = -\frac{5}{\sqrt{41}}\mathbf{i} - \frac{4}{\sqrt{41}}\mathbf{j}$$

The magnitute of the maximum rate of change is

$$\|\nabla h(10,4)\| = 4\sqrt{41}$$

In other words, the largest slope of the hill at the point $(10, 4)$ is $4\sqrt{41} \approx 25.6$ metres per kilometre of horizontal distance.

In the second question we are asked to identify all directions for which

$$D_{\mathbf{u}} f(10,4) > 0.7\,\|\nabla h(10,4)\|$$

Thus

$$\|\nabla h(10,4)\| \cos\theta > 0.7\,\|\nabla h(10,4)\|$$
$$\cos\theta > 0.7$$

Thus, the slope of the hill is larger than 70% of the maximum slope in all directions that make an angle θ, $0 \leq \theta < \arccos 0.7$, with respect to the gradient vector $\nabla h(10,4)$. Using a calculator, we find that $\cos 0.795 \approx 0.7$ (angle measure is in radians). So the answer is $0 \leq \theta < 0.795$.

Next, we study the geometry of the gradient vector in a bit more detail.

Pick a point (a, b) in the domain of a differentiable function f. Since f is differentiable, we know that its level curves near (a, b) are approximated by parallel lines. Draw the level curve of f that goes through (a, b); if $f(a,b) = c$, then the level curve has the value c; see Figure 9.8a.

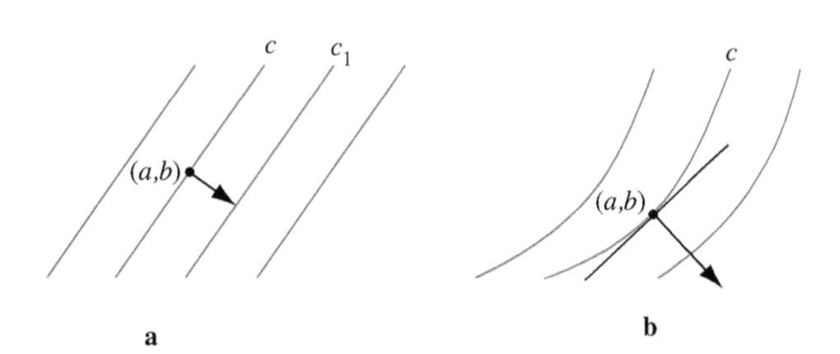

FIGURE 9.8
The gradient vector is perpendicular to a level curve

We look at different paths from (a, b) to a neighbouring level curve of a (slightly larger) value $c_1 > c$. The largest rate of change will be obtained by reaching the level curve of value c_1 along the shortest path, i.e., perpendicular to the two level curves.

This argument proves that the gradient $\nabla f(a, b)$ (i.e., the direction of the largest increase in f at (a, b)) is perpendicular to the level curve $f(x, y) = c$. To be precise, the vector $\nabla f(a, b)$ is perpendicular to the tangent to the level curve of value c at (a, b); see Figure 9.8b.

Because the tangent is perpendicular to the gradient vector, the rate of change in the direction of the tangent is zero. This makes sense: as we walk along the level curve $f(x, y) = c$, the values of f do not change (see Example 9.3).

Example 9.8 Geometric Interpretation of the Gradient Vector

Figure 9.9 shows level curves of a differentiable function $f(x, y)$.

(a) Explain why the vectors located at A and B cannot represent the gradient of f.

(b) Explain why the vectors located at C and D can represent the gradient of f.

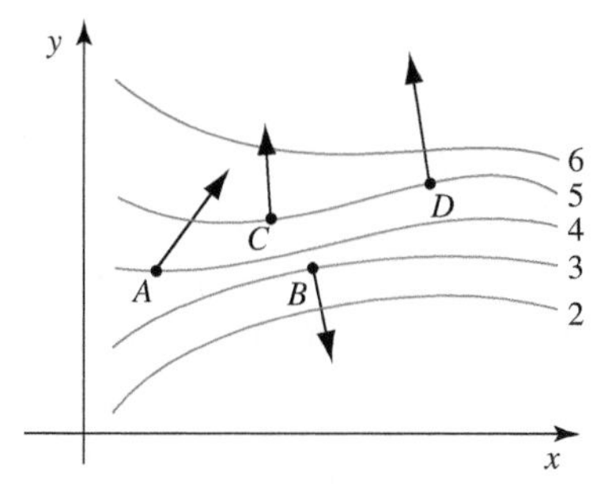

FIGURE 9.9
Geometric interpretation of the gradient vector

▶ The vector at A points toward increasing values of f but is not perpendicular to the level curve it sits on. The vector at B is perpendicular to the level curve but points in the wrong direction: toward decreasing rather than increasing values of f.

The vectors at C and D are perpendicular to the level curve and point in the direction of increasing values of f. Moreover, their relative size is correct: the level curves near D are closer, and so the rate of change at D is larger than the rate of change at C. ▲

To summarize, we list all the properties of the gradient vector that we covered in this section.

Theorem 16 Properties of the Gradient Vector

Assume that (a, b) is a point in the domain of a differentiable function f.

(a) If $\nabla f(a, b) = \mathbf{0}$, then the directional derivative $D_{\mathbf{u}} f(a, b) = 0$ in all directions.

(b) If $\nabla f(a, b) \neq \mathbf{0}$, then, at (a, b), f increases most rapidly in the direction of the vector $\nabla f(a, b)$; the magnitude of the most rapid increase is $\|\nabla f(a, b)\|$. The direction $-\nabla f(a, b)$ is the direction of the largest decrease in f at (a, b). The rate of change of f in that direction is $-\|\nabla f(a, b)\|$. In the directions perpendicular to $\nabla f(a, b)$, the directional derivative $D_{\mathbf{u}} f(a, b)$ is zero.

(c) If the gradient vector $\nabla f(a, b)$ is not zero, then it is perpendicular to the level curve of f that passes through (a, b).

Keep in mind that the directional derivative of a function makes sense only if all its independent variables have the same units. Otherwise, we can still calculate the directional derivative, but it has no meaning. For instance, the domain of a function $c(x, t)$, where x is distance (say, in metres) and t is time (in seconds), is a subset of a plane (xt-coordinate system) for which the horizontal axis represents the distance x and the vertical axis represents the time t. So far, so good—we can draw level curves, draw the graph of c, and so on.

However, concepts such as distance and vector do not have any meaning in such a coordinate system. The distance formula $\sqrt{(x_2 - x_1)^2 + (t_2 - t_1)^2}$ involves adding square metres and square seconds. As well, the components of a vector would have different units: how can we make sense of a vector with 10 m as one component and 25 s as the other?

Appendix: Vectors: Basic Facts and Formulas

We briefly review the definitions and results about vectors that we use in this section.

A *vector* $\mathbf{v}$ is an ordered pair $\mathbf{v} = (v_1, v_2)$ of real numbers called the *components* of $\mathbf{v}$. Often, instead of using the ordered pair notation, we write $\mathbf{v} = v_1\mathbf{i} + v_2\mathbf{j}$. The vectors $\mathbf{i} = (1, 0)$ and $\mathbf{j} = (0, 1)$ are *unit coordinate vectors.*

The vector $\mathbf{v}$ with initial point (tail) $A(a_1, a_2)$ and terminal point (head) $B(b_1, b_2)$, i.e., "the vector from $A(a_1, a_2)$ to $B(b_1, b_2)$," is given by

$$\mathbf{v} = (b_1 - a_1)\mathbf{i} + (b_2 - a_2)\mathbf{j}$$

Two vectors $\mathbf{v} = v_1\mathbf{i} + v_2\mathbf{j}$ and $\mathbf{w} = w_1\mathbf{i} + w_2\mathbf{j}$ are equal if and only if $v_1 = w_1$ and $v_2 = w_2$. The vector $\mathbf{0} = (0, 0)$, whose components are zero, is called the *zero vector.*

The *length* of a vector $\mathbf{v} = (v_1, v_2)$ is

$$\|\mathbf{v}\| = \sqrt{v_1^2 + v_2^2}$$

The length of the zero vector is zero. If $\mathbf{v} \neq \mathbf{0}$, then $\|\mathbf{v}\| > 0$. If $\mathbf{v}$ is the vector from $A(a_1, a_2)$ to $B(b_1, b_2)$, then $\mathbf{v} = (b_1 - a_1)\mathbf{i} + (b_2 - a_2)\mathbf{j}$ and its length

$$\|\mathbf{v}\| = \sqrt{(b_1 - a_1)^2 + (b_2 - a_2)^2}$$

is the distance between A and B.

A vector whose length is 1 is called a *unit vector.* For instance, $\mathbf{v} = 0.6\mathbf{i} + 0.8\mathbf{j}$ is a unit vector, since

$$\|\mathbf{v}\| = \sqrt{0.6^2 + 0.8^2} = \sqrt{0.36 + 0.64} = \sqrt{1} = 1$$

The *sum* of the vectors $\mathbf{v} = v_1\mathbf{i} + v_2\mathbf{j}$ and $\mathbf{w} = w_1\mathbf{i} + w_2\mathbf{j}$ is the vector

$$\mathbf{v} + \mathbf{w} = (v_1 + w_1)\mathbf{i} + (v_2 + w_2)\mathbf{j}$$

To multiply a vector $\mathbf{v} = v_1\mathbf{i}+v_2\mathbf{j}$ by a real number α, we multiply each component by α:

$$\alpha\mathbf{v} = \alpha v_1\mathbf{i} + \alpha v_2\mathbf{j}$$

The length of $\alpha\mathbf{v}$ is

$$\|\alpha\mathbf{v}\| = \sqrt{(\alpha v_1)^2 + (\alpha v_2)^2} = \sqrt{\alpha^2(v_1^2 + v_2^2)} = |\alpha|\sqrt{v_1^2 + v_2^2} = |\alpha|\,\|\mathbf{v}\|$$

The vector $\alpha\mathbf{v}$ points in the same direction as $\mathbf{v}$ if $\alpha > 0$, and in the opposite direction of $\mathbf{v}$ if $\alpha < 0$. If $\alpha = 0$, then $\alpha\mathbf{v} = \mathbf{0}$ (i.e., a zero vector).

Two non-zero vectors $\mathbf{v}$ and $\mathbf{w}$ are called *parallel* if there is a real number $\alpha \neq 0$ such that $\mathbf{w} = \alpha\mathbf{v}$.

Given a non-zero vector $\mathbf{v}$, the vector

$$\mathbf{u} = \frac{1}{\|\mathbf{v}\|}\mathbf{v} = \frac{\mathbf{v}}{\|\mathbf{v}\|}$$

points in the same direction as $\mathbf{v}$ (since $\|\mathbf{v}\| > 0$). Its length is

$$\|\mathbf{u}\| = \left\|\frac{1}{\|\mathbf{v}\|}\mathbf{v}\right\| = \left|\frac{1}{\|\mathbf{v}\|}\right|\,\|\mathbf{v}\| = \frac{1}{\|\mathbf{v}\|}\,\|\mathbf{v}\| = 1$$

so $\mathbf{u}$ is a unit vector. (Note that in this calculation we used the fact that $\|\alpha\mathbf{v}\| = |\alpha|\,\|\mathbf{v}\|$ with $\alpha = 1/\|\mathbf{v}\|$.) Thus, for any non-zero vector $\mathbf{v}$, the vector $\mathbf{u} = \mathbf{v}/\|\mathbf{v}\|$ is of unit length and points in the same direction as $\mathbf{v}$.

Example 9.9 Calculations with Vectors

The vector $\mathbf{v}$ from $A(3, -2)$ to $B(-4, 0)$ is given by $\mathbf{v} = (-4-3)\mathbf{i} + (0-(-2))\mathbf{j} = -7\mathbf{i} + 2\mathbf{j}$. Its length is

$$\|\mathbf{v}\| = \sqrt{(-7)^2 + (2)^2} = \sqrt{53}$$

The unit vector in the direction of $\mathbf{v} = -7\mathbf{i} + 2\mathbf{j}$ is

$$\mathbf{u} = \frac{-7\mathbf{i} + 2\mathbf{j}}{\|-7\mathbf{i} + 2\mathbf{j}\|} = \frac{-7\mathbf{i} + 2\mathbf{j}}{\sqrt{53}} = -\frac{7}{\sqrt{53}}\mathbf{i} + \frac{2}{\sqrt{53}}\mathbf{j}$$

The vector $3\mathbf{v} = 3(-7\mathbf{i} + 2\mathbf{j}) = -21\mathbf{i} + 6\mathbf{j}$ is three times as long as $\mathbf{v}$ and points in the same direction as $\mathbf{v}$. The vector $-\mathbf{v}/2 = (-7/2)\mathbf{i} + \mathbf{j}$ is half as long as $\mathbf{v}$ and points in the opposite direction of $\mathbf{v}$. Vectors $\mathbf{v}$, $3\mathbf{v}$, and $\mathbf{v}/2$ are parallel to each other.

The *angle* between vectors $\mathbf{v}$ and $\mathbf{w}$ is the smaller of the two angles formed by the directions of $\mathbf{v}$ and $\mathbf{w}$. So, if θ is the angle between two vectors, then $0 \leq \theta \leq \pi$. If $\theta = \pi/2$, we say that the vectors $\mathbf{v}$ and $\mathbf{w}$ are *orthogonal.*

The *dot product* $\mathbf{v} \cdot \mathbf{w}$ of vectors $\mathbf{v} = v_1\mathbf{i} + v_2\mathbf{j}$ and $\mathbf{w} = w_1\mathbf{i} + w_2\mathbf{j}$ is the real number

$$\mathbf{v} \cdot \mathbf{w} = v_1 w_1 + v_2 w_2$$

For example, if $\mathbf{v} = 3\mathbf{i} - \mathbf{j}$ and $\mathbf{w} = -2\mathbf{i} + 11\mathbf{j}$, then

$$\mathbf{v} \cdot \mathbf{w} = (3\mathbf{i} - \mathbf{j}) \cdot (-2\mathbf{i} + 11\mathbf{j}) = (3)(-2) + (-1)(11) = -17$$

As well,

$$\mathbf{v} \cdot \mathbf{i} = (3\mathbf{i} - \mathbf{j}) \cdot (\mathbf{i} + 0\mathbf{j}) = (3)(1) + (-1)(0) = 3,$$
$$\mathbf{i} \cdot \mathbf{i} = (1)(1) + (0)(0) = 1$$

and so on. Let $\mathbf{v} = v_1\mathbf{i} + v_2\mathbf{j}$; then

$$\mathbf{v} \cdot \mathbf{v} = (v_1\mathbf{i} + v_2\mathbf{j}) \cdot (v_1\mathbf{i} + v_2\mathbf{j}) = v_1 v_1 + v_2 v_2 = \|\mathbf{v}\|^2$$

gives a relationship between the dot product and the length. The dot product can be calculated in a geometric way. If $\mathbf{v} = v_1\mathbf{i} + v_2\mathbf{j}$ and $\mathbf{w} = w_1\mathbf{i} + w_2\mathbf{j}$, then

$$\mathbf{v} \cdot \mathbf{w} = \|\mathbf{v}\| \, \|\mathbf{w}\| \, \cos\theta \tag{9.5}$$

where θ is the angle between $\mathbf{v}$ and $\mathbf{w}$.

From (9.5) we conclude that if two *non-zero* vectors satisfy $\mathbf{v} \cdot \mathbf{w} = 0$, then $\cos\theta = 0$, i.e., $\theta = \pi/2$. As well, if $\theta = \pi/2$, then

$$\mathbf{v} \cdot \mathbf{w} = \|\mathbf{v}\| \, \|\mathbf{w}\| \, \cos\theta = \|\mathbf{v}\| \, \|\mathbf{w}\| \, 0 = 0$$

Thus, two non-zero vectors are orthogonal (or perpendicular) if and only if their dot product is zero.

For example, from

$$(3\mathbf{i} - \mathbf{j}) \cdot (4\mathbf{i} + 12\mathbf{j}) = (3)(4) + (-1)(12) = 0$$

it follows that the vectors $3\mathbf{i} - \mathbf{j}$ and $4\mathbf{i} + 12\mathbf{j}$ are orthogonal. Since $\mathbf{i}$ and $\mathbf{j}$ are orthogonal, their dot product $\mathbf{i} \cdot \mathbf{j}$ is zero.

All concepts that we have studied in this section can be extended to functions of three or more variables.

Summary By using the **directional derivative** we can calculate the rate of change of a function in any direction. Writing the partial derivatives of a function in the form of a vector, we obtain the **gradient vector.** The directional derivative is the dot product of the gradient vector and the unit vector in the desired direction. The gradient vector points in the direction of the **largest increase of a function,** and its magnitude is equal to that largest rate of change. In the direction opposite to the gradient, the function decreases the most. The gradient vector is **perpendicular to the level curves.**

9 Exercises

1. Write the limit definition of the directional derivative $D_{\mathbf{u}}f(a, b)$ in the case where $\mathbf{u} = \mathbf{i}/\sqrt{5} + 2\mathbf{j}/\sqrt{5}$, $f(x, y) = x^2 - y$, and $(a, b) = (2, -3)$.

2. Given two unit vectors $\mathbf{u}_1 = \mathbf{i}/\sqrt{2} + \mathbf{j}/\sqrt{2}$ and $\mathbf{u}_2 = \mathbf{i}/\sqrt{2} - \mathbf{j}/\sqrt{2}$, is it possible that, at the same point (a, b), $D_{\mathbf{u}_1}f(a, b) > 0$ and $D_{\mathbf{u}_2}f(a, b) > 0$? Explain why or why not.

3. Give a reason why $D_{\mathbf{u}}f(a, b) = 0$ if $\mathbf{u}$ is tangent to the contour curve of f passing through (a, b).

4. It is known that $\nabla f(2, 3) = 4\mathbf{i} - \mathbf{j}$. Is there a direction $\mathbf{u}$ such that $D_{\mathbf{u}}f(2, 3) = 5$? If so, find it.

5. It is known that $\nabla f(2, 3) = 4\mathbf{i} - \mathbf{j}$. Is there a direction $\mathbf{u}$ such that $D_{\mathbf{u}}f(2, 3) = 4$? If so, find it.

6. Let $f(x, y) = x^2y - 2y$ and $\mathbf{v} = 3\mathbf{i} - 4\mathbf{j}$.

   (a) Compute $D_{\mathbf{u}}f(1, 1)$, where $\mathbf{u}$ is the unit vector in the direction of $\mathbf{v}$.

   (b) Calculate $D_{\mathbf{v}}f(1, 1)$ for the vector $\mathbf{v}$ as given, ignoring the fact that we need a unit vector.

   (c) What is the relation between your answers to (a) and (b)? So why do we need to use a unit vector to calculate the directional derivative?

7. A curve is given implicitly by $F(x, y) = 0$, where F is a differentiable function. Explain how you can use the gradient to find the equation of the line tangent to the curve at the point (a, b).

8–13 ▪ Consider the contour diagram of a function $f(x, y)$.

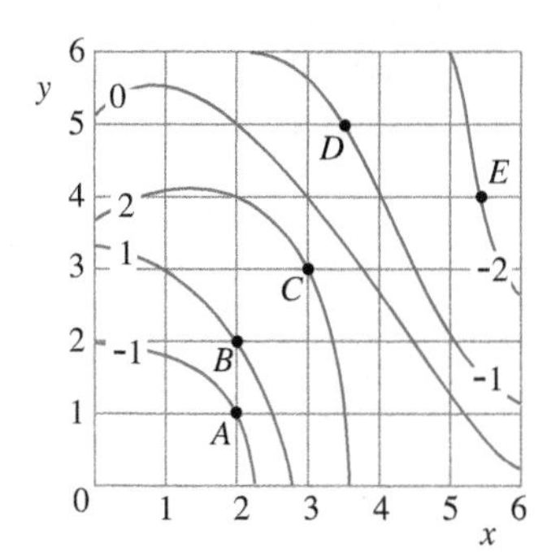

FIGURE 9.10

Estimate the value of the directional derivative $D_{\mathbf{u}}f$ at the given point in the given direction.

8. At A, in the direction $\mathbf{v} = \mathbf{j}$
9. At B, in the direction $\mathbf{v} = \mathbf{i} + \mathbf{j}$
10. At B, in the direction $\mathbf{v} = 3\mathbf{i} - 2\mathbf{j}$
11. At C, in the direction $\mathbf{v} = -\mathbf{i} + 2\mathbf{j}$
12. At D, in the direction $\mathbf{v} = -\mathbf{i}$
13. At E, in the direction $\mathbf{v} = -2\mathbf{i} + \mathbf{j}$

14. Draw gradient vectors at several points on the contour curve of value 0 in Figure 9.10.

15. Draw gradient vectors at several points on the contour curve of value 1 in Figure 9.10.

16–21 ▪ Find the directional derivative of the given function at the point A in the direction of the vector $\mathbf{v}$.

16. $f(x, y) = e^{xy}(\cos x + \sin x)$, $A = (\pi/2, 0)$, $\mathbf{v} = \mathbf{i} - \mathbf{j}$

17. $f(x, y) = x^3y + 2xy^2 - y^3$, $A = (1, 3)$, $\mathbf{v} = 2\mathbf{i} + \mathbf{j}$

18. $f(x, y) = x(x^2 + y^2)^{-1/2}$, $A = (1, 1)$, $\mathbf{v} = 2\mathbf{i} - 5\mathbf{j}$

19. $f(x, y) = \sqrt{x^2 + y^2}$, $A = (-2, -2)$, $\mathbf{v} = -\mathbf{i} + \mathbf{j}$

20. $f(x, y) = e^{-x^2-y^2}$, $A = (0, 1)$, $\mathbf{v} = \mathbf{i} + \mathbf{j}$

21. $f(x, y) = x \ln y^2 + 2y - 3$, $A = (3, 5)$, $\mathbf{v} = 8\mathbf{i} + 6\mathbf{j}$

22. In what direction at the point $(1, 2)$ is the directional derivative of $f(x, y) = 6xy$ equal to 4? Specify the direction as an angle with respect to the gradient of f at the given point.

23. In what directions at the point $(1, 2)$ is the directional derivative of $f(x, y) = 6xy$ smaller than 2? Specify the directions as angles with respect to the gradient of f at the given point.

24–27 ▪ Find the maximum rate of change of each function f and the direction in which it occurs.

24. $f(x, y) = 2ye^x + e^{-x}$, at $(0, 0)$
25. $f(x, y) = x^2y^{-2}$, at $(2, -1)$
26. $f(x, y) = \arctan(3y/x)$, at $(1, 1)$
27. $f(x, y) = y\sqrt{x} + y^3$, at $(1, 2)$

28. Assume that $f(x, y)$ is a differentiable function. Identify all directions at the point (a, b) in which the rate of increase of f is at least 80% of the largest possible increase at that point.

29. Look at the contour diagram of f in Figure 9.10. Which of $\|\nabla f(A)\|$ or $\|\nabla f(B)\|$ is larger?

30. Look at the contour diagram of f in Figure 9.10. Which of $\|\nabla f(B)\|$ or $\|\nabla f(D)\|$ is larger?

31. Find $\nabla f(x, y)$ for $f(x, y) = \sqrt{x^2 + y^2}$. Give a geometric interpretation of f. Sketch or describe the gradient vectors $\nabla f(x, y)$.

32. Find $\nabla f(x,y)$ for $f(x,y) = x^2 + y^2$. Give a geometric interpretation of f. Sketch or describe the gradient vectors $\nabla f(x,y)$.

33. Suppose that we are looking at the contour diagram of a differentiable function f and determine the following: if we move from $(1,3)$ toward $(2,2)$, the directional derivative of f is 10; if we move from $(1,3)$ toward $(2,3)$, the directional derivative of f is 6. What is the gradient of f at $(1,3)$?

34. Someone has calculated the following for a differentiable function f: in the direction from $(1,1)$ toward $(2,2)$, the directional derivative is 10; in the direction from $(1,1)$ toward $(0,0)$, the directional derivative is 5. How do you know that there is an error in the calculation?

35. The pressure $P(x,y)$ at a point $(x,y) \in \mathbb{R}^2$ on a membrane is given by the function $P(x,y) = 40e^{-2x^2-y^2}$.

    (a) Find the rate of change of the pressure at the point $(1,0)$ in the direction $\mathbf{i} - \mathbf{j}$.

    (b) In what direction away from the point $(1,0)$ does the pressure increase most rapidly? Decrease most rapidly?

    (c) Find the maximum rate of increase of pressure at $(1,0)$.

    (d) Locate the direction(s) at $(1,0)$ in which the rate of change of pressure is zero.

36. The temperature produced by a source located at the origin is given by $T(x,y) = 12e^{-x^2-y^2}$.

    (a) Sketch the isothermal curves, i.e., the curves on which the temperature is constant. Find the gradient ∇T and add several gradient vectors to your sketch.

    (b) Which point is the warmest?

    (c) With (a) and (b) in mind, sketch (or describe) the gradient vector field of T.

    (d) What is the direction of the most rapid decrease in temperature at the point $(1,2)$? What is the magnitude of that decrease?

37. The temperature at a point (x,y) in a plane is given by $T(x,y) = 120(x^2 + y^2 + 4)^{-1}$.

    (a) What do the level curves of T look like? Make a sketch of a contour diagram of T.

    (b) Calculate ∇T and add several gradient vectors to your sketch.

    (c) Based on your diagram in (b), can you determine which is the warmest point?

    (d) Find the warmest point algebraically and compare with your answer to (c).

38. The direction of the greatest increase in height at some point on a hill, 25 m per 100 m of horizontal distance, is toward the west. At this point, in the direction toward the southwest, how steep is the hill?

39. Find an equation of the line tangent to the curve given implicitly by the equation $e^y - 2xy^3 + 4x = 5$ at $(1,0)$.

40. A curve is given implicitly by $y^2 \sin x - xy + 2x = \pi$. Find an equation of the tangent to the curve at $(\pi/2, 0)$.

10 Extreme Values

Having learned about derivatives and how they describe change, we are now ready to investigate **extreme values** of functions of several variables. As in the case of functions of one variable, we study **relative (local)** and **absolute (global)** extreme values. However, two major differences make our investigations quite involved at times: we now have two (and not one) first partial derivatives and three or four (and not one) second partial derivatives; as well, we work with subsets of the plane (and not with intervals).

Local Extreme Values

Recall that a function $y = f(x)$ has a local maximum at $x = a$ if $f(a) \geq f(x)$ for all x near a. The phrase "for all x near a" means for all x in some open interval around a (Figure 10.1a).

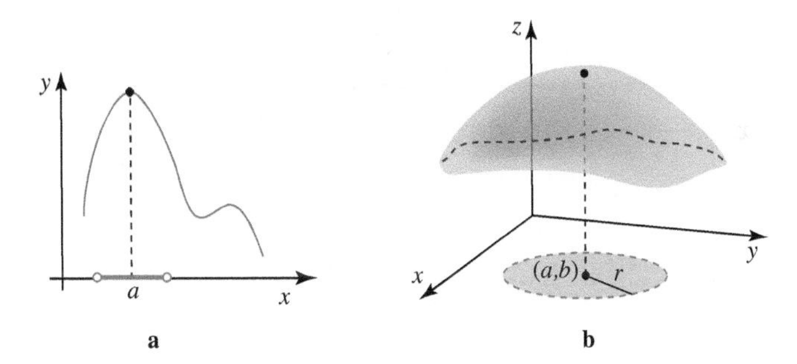

FIGURE 10.1
The meaning of "near"

In the case of two variables, "for all (x, y) near (a, b)" means for all points (x, y) in some open disk $B_r(a, b)$ centred at (a, b) (Figure 10.1b). Recall that the open disk $B_r(a, b)$ is the inside of the disk of radius r centred at (a, b):

$$B_r(a,b) = \{(x,y) \in \mathbb{R}^2 \mid \sqrt{(x-a)^2 + (y-b)^2} < r\}$$

Definition 18 Local Extreme Values

We say that a function $f(x, y)$ has a *local maximum* at a point (a, b) in its domain if

$$f(a,b) \geq f(x,y)$$

for all (x, y) near (a, b). If

$$f(a,b) \leq f(x,y)$$

for all (x, y) near (a, b), then f has a *local minimum* at (a, b).

Local maxima and local minima are referred to as *local extremes* or *local extreme values.*

A local extreme value requires two pieces of information: where it happens (the point (a, b) in the domain of f) and what its value is (the real number $f(a, b)$). When we wish to be precise, we will say that (a, b) is a local maximum (minimum, extreme) *point*, and $f(a, b)$ is a local maximum (minimum, extreme) *value,* or we will say that f has a local maximum (minimum, extreme) at (a, b). However, when there is no danger of confusion, we will often just say that (a, b) is a local maximum (minimum, extreme) of f.

Example 10.1 Functions and Extreme Values

Consider the function $f(x, y) = 2x^2 + 3y^2 + 5$. Since $x^2 \geq 0$ and $y^2 \geq 0$ for all x and y, it follows that

$$f(x, y) \geq 5$$

for all (x, y) in $\mathbb{R}^2$. As well, $f(0, 0) = 5$. Thus, f has a local minimum at $(0, 0)$, and its local minimum value is $f(0, 0) = 5$. Referring to Definition 18, in this case we can take any open disk $B_r(0, 0)$ to specify how near (how close) a point (x, y) should be to $(0, 0)$.

Since f is an increasing function of x and y, it has no local maximum. In Figure 10.2a we show the graph of $f(x, y)$, and in Figure 10.2b we show its level curves.

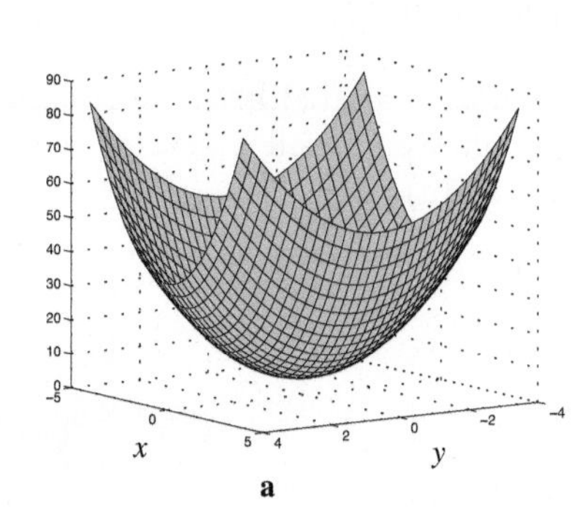

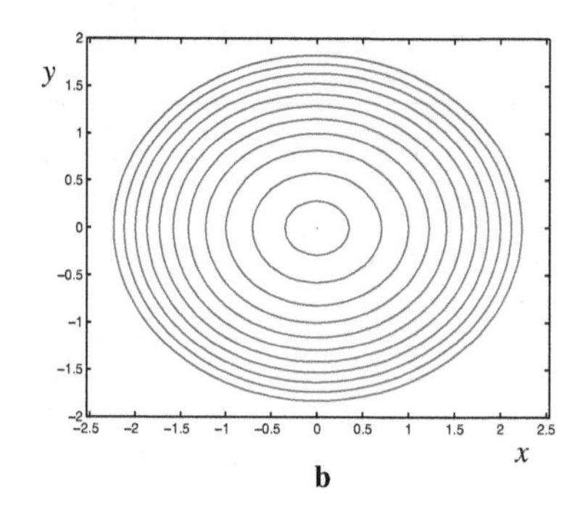

FIGURE 10.2

Graph and level curves of $f(x, y) = 2x^2 + 3y^2 + 5$

Next, look at the function $g(x, y) = e^{-x^2-y^2}$. From

$$-x^2 - y^2 = -(x^2 + y^2) \leq 0$$

we conclude that

$$e^{-x^2-y^2} \leq e^0 = 1$$

for all (x, y). In other words, $g(x, y)$ is either less than 1 or equal to 1 for all (x, y). Since $g(0, 0) = 1$, it follows that g has a local maximum at $(0, 0)$, and the local maximum value is $g(0, 0) = 1$; see Figure 10.3.

Note that $e^{-x^2-y^2} > 0$ for all x and y, and $e^{-x^2-y^2} \to 0$ as $x \to \pm\infty$ or $y \to \pm\infty$ (or both). Thus, $g(x, y)$ has no local minimum.

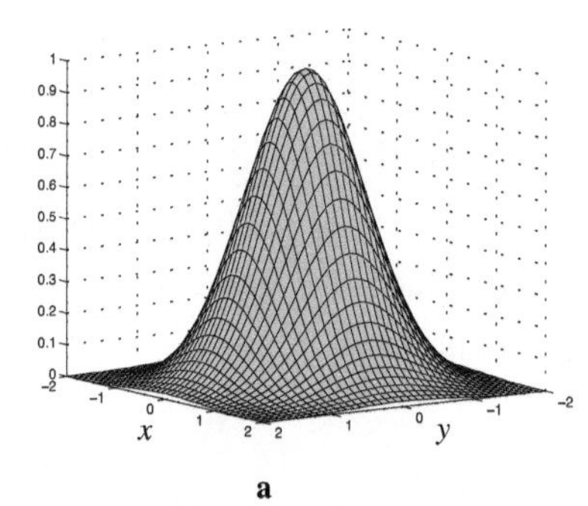

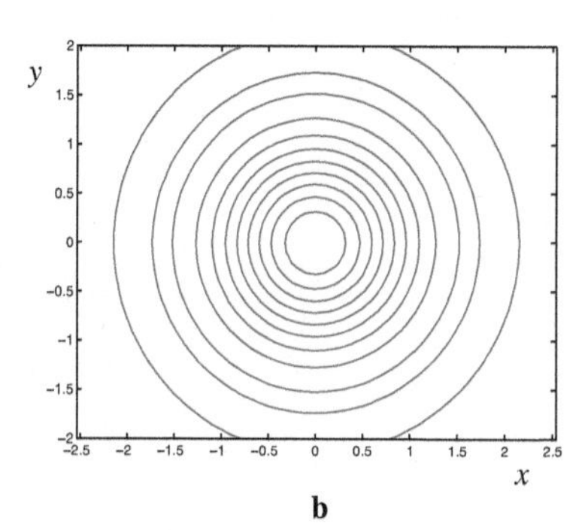

FIGURE 10.3

Graph and level curves of $g(x, y) = e^{-x^2-y^2}$

Since $-1 \leq \sin(x + y) \leq 1$ for all (x, y) in $\mathbb{R}^2$, the function $h(x, y) = \sin(x + y)$ has a local minimum value of -1, which occurs at all points (x, y) such that $x + y = -\pi/2 + 2\pi k$ (k is an integer). The local maximum of h is 1, occurring at all points (x, y) such that $x + y = \pi/2 + 2\pi k$.

Figure 10.4 shows the graph of h as well as its level curves.

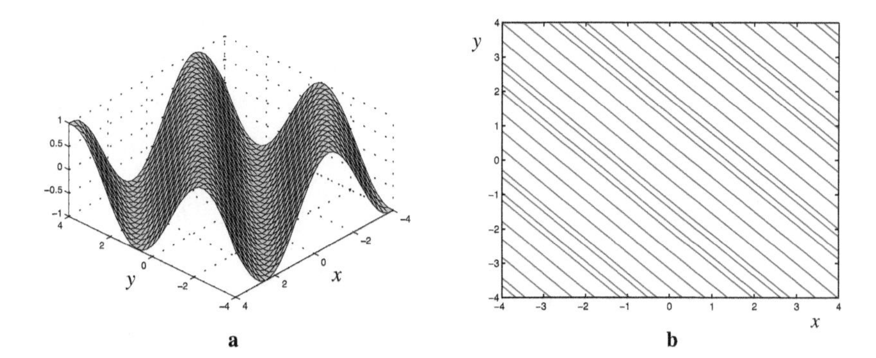

FIGURE 10.4
Graph and level curves of $h(x, y) = \sin(x + y)$

Pick a point for which $x + y = \pi/2 + 2\pi k$; for instance, take $k = 0$, and then pick $x = \pi/2$ and $y = 0$. The function h has a local maximum at $(\pi/2, 0)$:

$$h(\pi/2, 0) = \sin(\pi/2 + 0) = 1$$

Note that any open disk centred at $(\pi/2, 0)$ contains points that lie on the line $x + y = \pi/2$, at which h is equal to 1; see Figure 10.5.

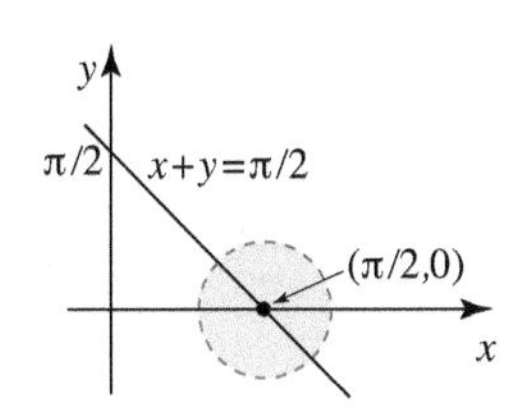

FIGURE 10.5
Analyzing the local maximum of $h(x, y) = \sin(x + y)$

This does not violate the definition of local maximum—Definition 18 states that $f(a, b)$ is a local maximum if there are *no larger values near (a,b).* In other words, what is unique is the local maximum *value,* and not the location where it occurs. (Of course, the same is true for a local minimum.)

Example 10.2 A Function with No Local Extreme Values

The linear function $l(x, y) = x + y$ has no local extreme values.

How do we know? We add two numbers x and y and get $x + y$. If we add x and a number a bit larger than y, we get more than $x + y$. So, $x + y$ cannot be a maximum. Adding x and a number a bit smaller than y, we obtain a number smaller than $x + y$, so $x + y$ cannot be a minimum either.

Let's formalize this thinking. Pick a point (a, b) and an open disk $B_r(a, b)$. We show that $l(x, y)$ cannot have a maximum or a minimum at (a, b).

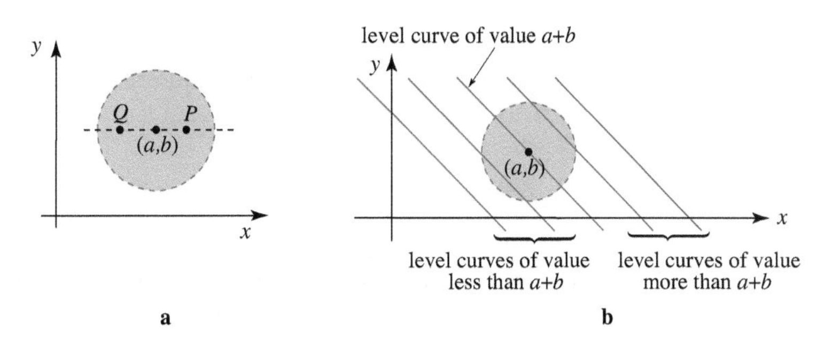

FIGURE 10.6
Analyzing $l(x, y) = x + y$

The value of l at (a, b) is $l(a, b) = a + b$. Inside $B_r(a, b)$, pick a point P to the right of (a, b) and a point Q to the left of it; see Figure 10.6a. The coordinates of P are

$(a + a_1, b)$, where $a_1 > 0$, and the coordinates of Q are $(a - a_2, b)$, where $a_2 > 0$. It follows that

$$l(P) = f(a + a_1, b) = a + a_1 + b > a + b = l(a, b)$$

so l does not have a maximum at (a, b). It does not have a minimum at (a, b) either, since

$$l(Q) = f(a - a_2, b) = a - a_2 + b < a + b = l(a, b)$$

Alternatively, look at the level curves of $l(x, y)$ in Figure 10.6b. Any open disk centred at (a, b) contains level curves that lie on both sides of the level curve of value $a + b$, where the function l is either smaller than or greater than $a + b$. ▲

As Examples 10.1 and 10.2 show, a function may or may not have local extreme values, or it may have a minimum only or a maximum only. How do we figure out whether a given function has local extreme values, and, if it does, how do we find them?

Assume that $f(x, y)$ is differentiable. If, at some point (a, b), the gradient $\nabla f(a, b)$ is not zero, then $f(a, b)$ is not an extreme value—if we move in the direction of the gradient, the values of f increase (and so $f(a, b)$ is not a maximum). If we move in the opposite direction, the values of f decrease (and so $f(a, b)$ is not a minimum).

Thus, the only points (a, b) where a differentiable function f might have a local minimum or a local maximum are those where the gradient vector $\nabla f(a, b)$ is zero, i.e., where $f_x(a, b) = 0$ and $f_y(a, b) = 0$.

To illustrate this fact, Figure 10.7 shows gradient vectors drawn at various points in the domains of the functions $f(x, y) = 2x^2 + 3y^2 + 5$ and $g(x, y) = e^{-x^2-y^2}$ of Example 10.1. The gradient vectors of f point away from $(0, 0)$, indicating that moving away from $(0, 0)$ increases the values of f (indeed, $(0, 0)$ is a local minimum). The gradient vectors of g point toward $(0, 0)$; thus, moving toward $(0, 0)$ increases the values of g, which suggests that g has a maximum there.

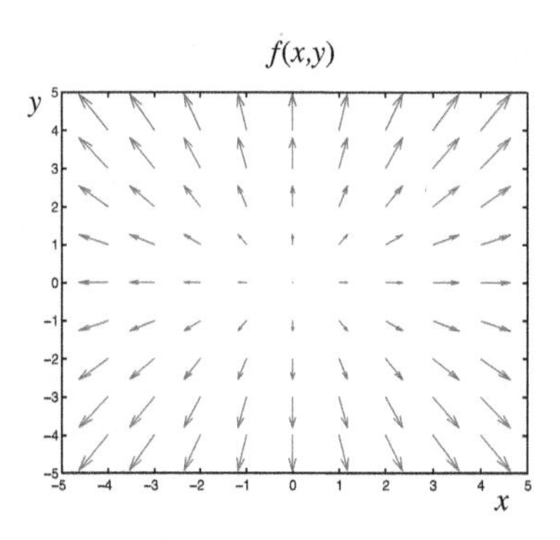

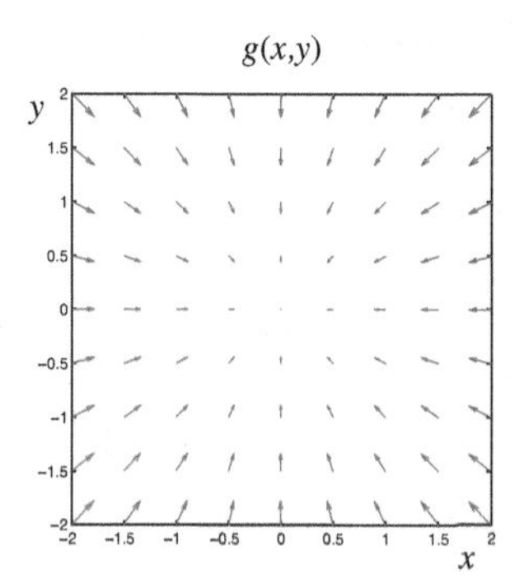

FIGURE 10.7

Gradient vectors of $f(x, y) = 2x^2 + 3y^2 + 5$ and $g(x, y) = e^{-x^2-y^2}$

So, at a point (a, b) where a differentiable function f has an extreme value, $f_x(a, b) = f_y(a, b) = 0$. We give an alternative proof of this fact that does not use the gradient.

Assume that $f(a, b)$ is a local maximum, and define the function of *one* variable

$$g(x) = f(x, b)$$

The graph of g is the intersection of the surface that represents the graph of $z = f(x, y)$ and the vertical plane $y = b$ (we have already met this situation in Sections 4 and 5).

Definition 20 **Saddle Point**

A critical point that is neither a local minimum point nor a local maximum point is called a *saddle point.*

Next, we state the test that will help us determine whether or not a function f of two variables has an extreme value at a critical point. Although the test does not apply to all cases, it is nevertheless quite useful.

Analogous to the one-variable case, we will need to use the second derivatives of f. A way to keep them organized is to use a table (called a *matrix*):

$$\begin{bmatrix} f_{xx}(x,y) & f_{xy}(x,y) \\ f_{yx}(x,y) & f_{yy}(x,y) \end{bmatrix} \tag{10.1}$$

We assume that the second partial derivatives are continuous, so that $f_{xy}(x,y) = f_{yx}(x,y)$ (see Theorem 11 in Section 7).

The matrix (10.1) is called the *Hessian matrix of* f or just the *Hessian of* f. We will not work with the matrix but rather with the function $D(x,y)$ that we extract from it:

$$D(x,y) = f_{xx}(x,y)f_{yy}(x,y) - (f_{xy}(x,y))^2$$

Those familiar with matrices will recognize D as the determinant of the Hessian matrix in the case where the mixed partials are equal.

Theorem 19 **Second Derivatives Test**

Assume that all second-order partial derivatives of a function $f(x,y)$ are continuous on an open disk centred at (a,b) and that (a,b) is a critical point of f (i.e., $f_x(a,b) = 0$ and $f_y(a,b) = 0$).

(a) If $D(a,b) > 0$ and $f_{xx}(a,b) < 0$, then f has a local maximum at (a,b)

(b) If $D(a,b) > 0$ and $f_{xx}(a,b) > 0$, then f has a local minimum at (a,b)

(c) If $D(a,b) < 0$, then f has neither a local minimum nor a local maximum at (a,b).

Note that the assumption of f guarantees that $f_{xy}(x,y) = f_{yx}(x,y)$. If (c) holds, then the critical point (a,b) is a saddle point.

What happens in the cases that are not covered by Theorem 19?

If $D(a,b) = 0$, then the second derivatives test provides no answer: f could have a minimum, a maximum, or a saddle point at (a,b). In Example 10.9 we show how to analyze critical points in that case.

If $D(a,b) > 0$ and $f_{xx}(a,b) = 0$ but $f_{yy}(a,b) \neq 0$, then in parts (a) and (b) of Theorem 19 we use $f_{yy}(a,b)$ instead of $f_{xx}(a,b)$. If $f_{xx}(a,b) = f_{yy}(a,b) = 0$ but $f_{xy}(a,b) \neq 0$, then

$$D(x,y) = f_{xx}(x,y)f_{yy}(x,y) - (f_{xy}(x,y))^2 = -(f_{xy}(x,y))^2 < 0$$

and case (c) applies.

We now demonstrate how Theorem 19 is used. A sketch of the proof is given at the end of this section.

Example 10.8 Using the Second Derivatives Test

In Example 10.4 we discovered that the function $f(x, y) = x^4 + y^2 - 4xy + 2$ has three critical points: $(0, 0)$, $(\sqrt{2}, 2\sqrt{2})$, and $(-\sqrt{2}, -2\sqrt{2})$. Now we use the second derivatives test to determine what they represent.

▶ From $f_x = 4x^3 - 4y$ we compute $f_{xx} = 12x^2$ and $f_{xy} = -4$. From $f_y = 2y - 4x$ we compute $f_{yy} = 2$ and $f_{yx} = -4$ (same as f_{xy}). Thus,

$$D(x, y) = f_{xx}f_{yy} - (f_{xy})^2 = (12x^2)(2) - (-4)^2 = 24x^2 - 16$$

The fact that $D(0, 0) = -16 < 0$ implies that $(0, 0)$ is a saddle point. From

$$D(\sqrt{2}, 2\sqrt{2}) = 24\left(\sqrt{2}\right)^2 - 16 = 32 > 0$$

and

$$f_{xx}(\sqrt{2}, 2\sqrt{2}) = 12\left(\sqrt{2}\right)^2 = 24 > 0$$

we conclude that

$$\begin{aligned} f(\sqrt{2}, 2\sqrt{2}) &= \left(\sqrt{2}\right)^4 + \left(2\sqrt{2}\right)^2 - 4\left(\sqrt{2}\right)\left(2\sqrt{2}\right) + 2 \\ &= 4 + 8 - 16 + 2 = -2 \end{aligned}$$

is a local minimum. Likewise,

$$D(-\sqrt{2}, -2\sqrt{2}) = 24\left(-\sqrt{2}\right)^2 - 16 = 32 > 0$$

and

$$f_{xx}(-\sqrt{2}, -2\sqrt{2}) = 12\left(-\sqrt{2}\right)^2 = 24 > 0$$

imply that

$$f(-\sqrt{2}, -2\sqrt{2}) = \left(-\sqrt{2}\right)^4 + \left(-2\sqrt{2}\right)^2 - 4\left(-\sqrt{2}\right)\left(-2\sqrt{2}\right) + 2 = -2$$

is a local minimum. ▲

Example 10.9 A Case Where the Second Derivatives Test Gives No Answer

Analyze the function $f(x, y) = x^3 - 3xy^2$ for the existence of local extreme values.

▶ From $f_x = 3x^2 - 3y^2 = 0$ it follows that $y^2 = x^2$ and $y = \pm x$. From $f_y = -6xy = 0$ we get $xy = 0$.

Substituting $y = x$ into $xy = 0$ yields $x^2 = 0$ and $x = 0$. In that case, $y = 0$ as well, and so $(0, 0)$ is a critical point. Substituting $y = -x$ into $xy = 0$ yields $-x^2 = 0$ and $x = 0$. Again, we obtain $(0, 0)$. Since f_x and f_y are defined for all (x, y), we conclude that $(0, 0)$ is the only critical point of f.

For the second derivatives test, we compute $f_{xx} = 6x$, $f_{xy} = f_{yx} = -6y$, $f_{yy} = -6x$, and

$$D(x, y) = f_{xx}f_{yy} - (f_{xy})^2 = (6x)(-6x) - (-6y)^2 = -36x^2 - 36y^2$$

Since $D(0, 0) = 0$, Theorem 19 provides no answer. So what is $(0, 0)$?

Note that $f(0, 0) = 0$. If we take $y = 0$, then $f(x, 0) = x^3$. So, near $(0, 0)$, f takes on both positive and negative values, which means that $f(0, 0)$ cannot be a local extreme value. Thus, it must be a saddle point; see Figure 10.12.

This surface is sometimes called a *monkey saddle*. Unlike the usual saddle surface (Figure 10.11), this one is modelled for a rider with a tail. ▲

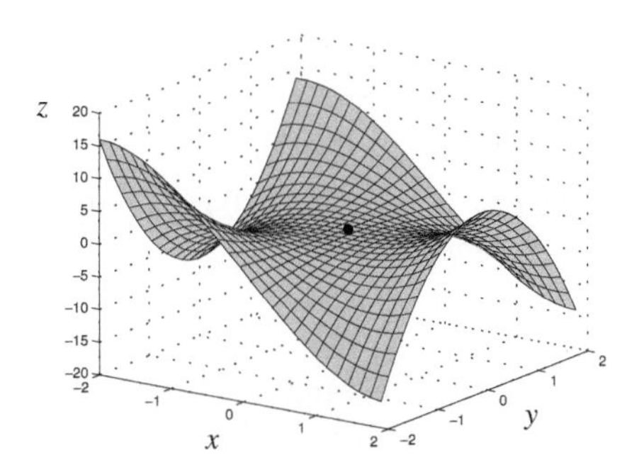

FIGURE 10.12

The graph of $f(x, y) = x^3 - 3xy^2$

In Exercises 35 and 36, we explore more examples of functions for which $D = 0$, i.e., for which the second derivatives test does not apply.

Example 10.10 Building an Optimal Box

Find the dimensions of the rectangular box with a volume of 64 cm^3 and the smallest possible surface area. (In other words, we are asked to build a rectangular box (with a lid) that will hold the required volume, using the least amount of material.)

Let x, y, and z denote the length, width, and height of the box; hence $x, y, z > 0$. The assumption states that $xyz = 64$. We are asked to minimize the function of three variables

$$A = 2xy + 2yz + 2xz$$

To write this as a function of two variables, we eliminate z: from $xyz = 64$ it follows that $z = 64/xy$ and thus

$$A(x, y) = 2xy + 2y\frac{64}{xy} + 2x\frac{64}{xy} = 2xy + \frac{128}{x} + \frac{128}{y}$$

Now we apply the usual routine of searching for local extreme values. From

$$A_x = 2y - \frac{128}{x^2} = 0$$

we get $2y = 128/x^2$ and $y = 64/x^2$. Likewise, from

$$A_y = 2x - \frac{128}{y^2} = 0$$

we get $x = 64/y^2$. Combining the two equations we get

$$x = \frac{64}{y^2} = \frac{64}{(64/x^2)^2} = \frac{64}{64^2/x^4} = \frac{x^4}{64}$$

and thus

$$x - \frac{x^4}{64} = 0$$

$$x\left(1 - \frac{x^3}{64}\right) = 0$$

It follows that $x = 0$ (not in the domain of A) and $x^3/64 = 1$, i.e., $x = 4$. Now $y = 64/x^2 = 64/16 = 4$; thus, $(4, 4)$ is the only critical point.

To check that $(4, 4)$ is indeed a minimum, we apply the second derivatives test. From $A_{xx} = 256/x^3$, $A_{xy} = A_{yx} = 2$, and $A_{yy} = 256/y^3$, we compute

$$D = A_{xx}A_{yy} - (A_{xy})^2 = \frac{256}{x^3}\frac{256}{y^3} - (2)^2 = \frac{256^2}{x^3y^3} - 4$$

Now

$$D(4,4) = \frac{256^2}{(64)(64)} - 4 = 16 - 4 = 12 > 0$$

and $A_{xx}(4,4) = 256/4^3 > 0$ imply that $(4,4)$ is indeed a minimum.

The height of the box is $z = 64/xy = 64/16 = 4$; thus, the box has the shape of a cube of side length 4. ▲

We showed that, of all rectangular boxes, the optimal box (i.e., the one requiring the least amount of material to be built) is the one that is the most symmetric: the cube.

If we remove the condition that the box must be rectangular, and consider all possible shapes instead, it would turn out, again, that the optimal shape is the one that is the most symmetric: the sphere. (That is not easy to prove.)

There is a major principle behind this. Optimization—in nature, and elsewhere—forces symmetry. For instance, an animal sleeping in the snow curls up. Why? Since it cannot make itself smaller (i.e., cannot change its volume), the animal tries to minimize its surface area (knowing that heat loss is proportional to surface area). The closer it gets to assuming a spherical shape, the less heat it will lose.

See Example 10.14, in which we revisit the optimal box problem.

Example 10.11 Regression Line

When we investigate a phenomenon, or conduct an experiment, we collect data. In some cases, those data can be represented in a plane, as in Figure 10.13a. In order to use calculus to analyze data, we need functions. How do we move from data to functions? In other words, how do we find the function that best matches our data (Figure 10.13b)?

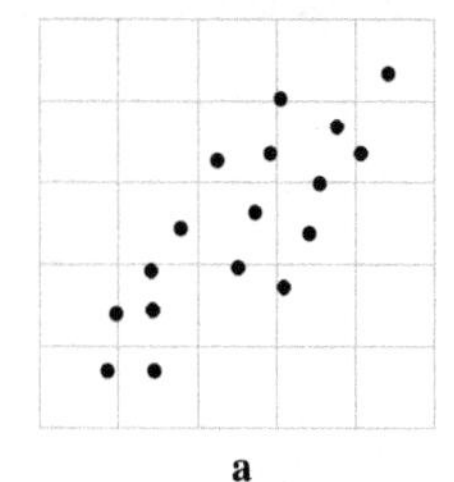

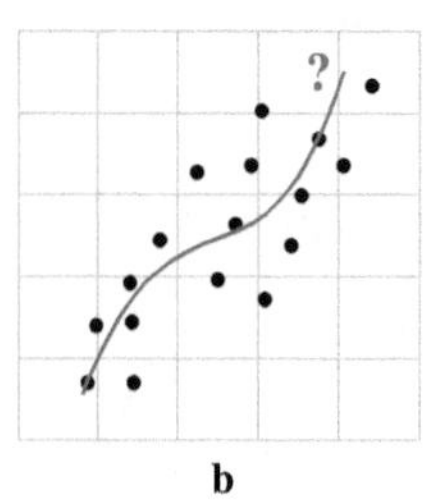

FIGURE 10.13

Matching a function to given data

Of numerous approaches and techniques used to identify the function that best matches given data, we outline the *least squares line* or the *regression line* method.

There is a big difference between the two sets of data in Figure 10.14. In Figure 10.14a, the data are scattered all over the plane, whereas in Figure 10.14b they seem to cluster along a line. When we have data as in Figure 10.14b, it might be a good idea to find the linear function (line) that best describes it.

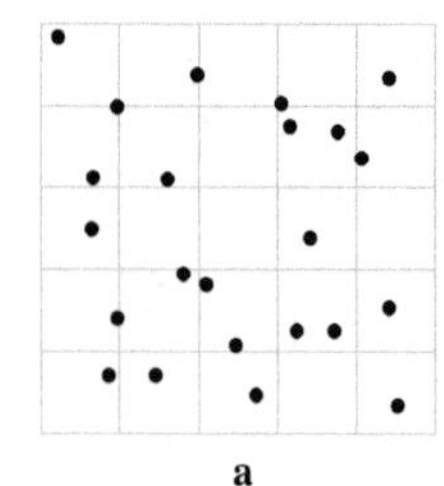

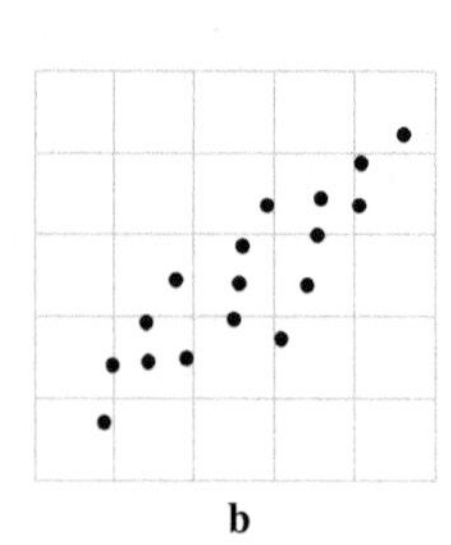

FIGURE 10.14

Different data sets

We now show how to do this. To simplify calculations, we use three points (the general case of n points is not difficult, just messier; see Exercise 45).

Assume that we are given data points $(2,1)$, $(3,1)$, and $(4,3)$. We are looking for the line $y = mx + b$ that best fits the three given points in the following sense: we will compute the sum of the squares of the vertical distances (see Figure 10.15a) between the data points and the line $y = mx + b$ and then minimize that sum. In doing so, we will find the values of m and b.

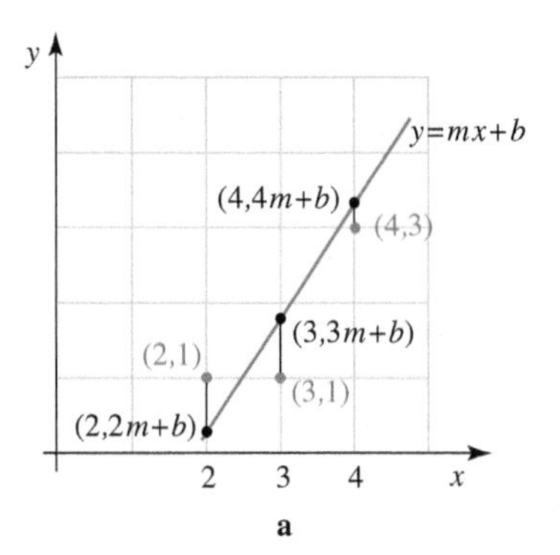

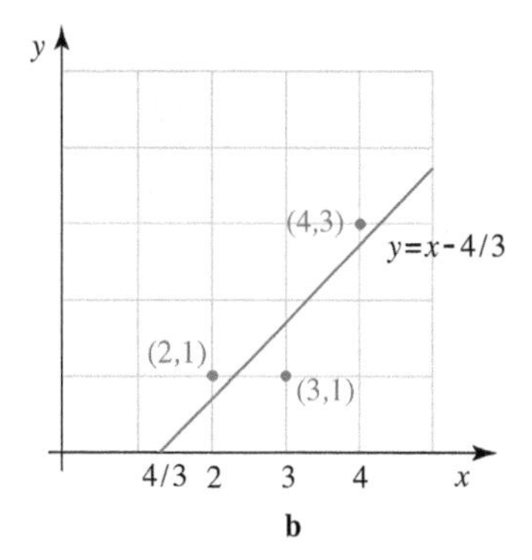

FIGURE 10.15

Finding the line that best fits the three points

The vertical distance between $(2,1)$ and the corresponding point $(2, 2m+b)$ on the line is $|(2m+b)-1|$. The vertical distance between $(3,1)$ and the corresponding point $(3, 3m+b)$ on the line is given by $|(3m+b)-1|$. The remaining distance is $|(4m+b)-3|$, and the function we need to minimize is

$$f(m,b) = ((2m+b)-1)^2 + ((3m+b)-1)^2 + ((4m+b)-3)^2$$

The partial derivatives of f are

$$\begin{aligned} f_m &= 2((2m+b)-1)(2) + 2((3m+b)-1)(3) + 2((4m+b)-3)(4) \\ &= 8m + 4b - 4 + 18m + 6b - 6 + 32m + 8b - 24 \\ &= 58m + 18b - 34 \end{aligned}$$

and

$$\begin{aligned} f_b &= 2((2m+b)-1) + 2((3m+b)-1) + 2((4m+b)-3) \\ &= 4m + 2b - 2 + 6m + 2b - 2 + 8m + 2b - 6 \\ &= 18m + 6b - 10 \end{aligned}$$

To find critical points, we solve the system (divide both $f_m = 0$ and $f_b = 0$ by 2):

$$\begin{aligned} 29m + 9b - 17 &= 0 \\ 9m + 3b - 5 &= 0 \end{aligned}$$

Multyplying the second equation by -3 and adding to the first, we get $2m - 2 = 0$ and $m = 1$. Substituting $m = 1$ into the second equation, $9(1) + 3b - 5 = 0$, i.e., $b = -4/3$. Thus, there is only one critical point, $(m = 2, b = -4/3)$.

To show that this critical point yields a minimum, we apply the second derivatives test. From $f_{mm} = 58$, $f_{mb} = f_{bm} = 18$, and $f_{bb} = 6$, we compute

$$D = f_{mm} f_{bb} - f_{mb}^2 = (58)(6) - (18)^2 = 24 > 0$$

Since $f_{mm} = 58 > 0$, the point $(2, -4/3)$ is indeed a minimum. Thus, the line that best fits the given data (in the sense explained at the start) is $y = x - 4/3$; see Figure 10.15b. ▲

Absolute Extreme Values

A function $y = f(x)$ of one variable is said to have an absolute maximum (minimum) on a set S if there is a number a in S such that $f(a) \geq f(x)$ ($f(a) \leq f(x)$) for *all* x in S. Recall that there is one situation where absolute extreme values are guaranteed to exist.

Theorem 20 Extreme Value Theorem for Functions of One Variable

Assume that $y = f(x)$ is a continuous function defined on a closed interval $[c, d]$. Then f has an absolute maximum and an absolute minimum at some points a_1 and a_2 in $[c, d]$.

We would like to generalize this statement to functions of two variables. Theorem 20 has two assumptions: the function needs to be continuous and the interval where it is defined must be closed. We know how to define continuity for functions of two variables, but we need to explain what the two-dimensional analogue of a closed interval is.

What makes a closed interval $[c, d]$ "closed" is the fact that it contains its boundary points c and d. And there is one more thing; since c and d are real numbers (and not $-\infty$ or ∞), the interval $[c, d]$ is bounded, i.e., is of finite length (unlike $(-\infty, 0]$ or $[3, \infty)$).

So, to extend Theorem 20 to functions of two variables we will replace the assumption "closed interval" by the assumption "closed and bounded set." First, we explain what exactly closed and bounded sets in $\mathbb{R}^2$ are.

Definition 21 Boundary Point

A point P is called a *boundary point* of a set S in $\mathbb{R}^2$ if every open disk B_r centred at P contains points in S and also points not in S.

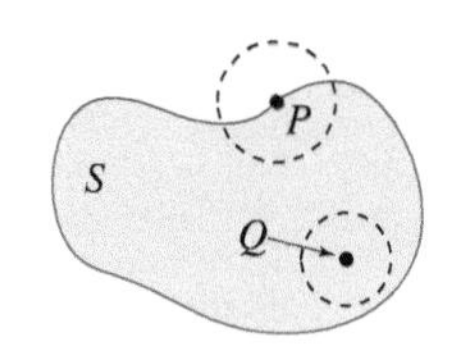

FIGURE 10.16
Boundary point

The point P in Figure 10.16 is a boundary point of the set S. An open disk, no matter how small, contains points that lie both inside and outside S. On the other hand, the point Q is not a boundary point. The disk centred at Q contains no points outside S. (Note that Definition 21 requires that *every* open disk contain points both from the inside and from the outside of S.)

So the concept of a boundary point, at least in the case of $\mathbb{R}^2$ that we consider here, agrees with our intuitive notion of what a boundary point should be.

Definition 22 Closed Set

A set S in $\mathbb{R}^2$ is *closed* if it contains all of its boundary points.

Recall the convention we use in drawing: a solid line or a solid curve represents points that belong to the set, whereas a dashed line or a dashed curve represents points that do not belong to the set.

The disk $D = \{(x, y) \mid x^2 + y^2 \leq 1\}$ is closed: all of its boundary points lie on the circle $\{(x, y) \mid x^2 + y^2 = 1\}$, which is a part of D; see Figure 10.17a. The first quadrant $S = \{(x, y) \mid x \geq 0 \text{ and } y \geq 0\}$ is closed because it contains all of its boundary points (the origin, the positive x-axis, and the positive y-axis); see Figure 10.17b.

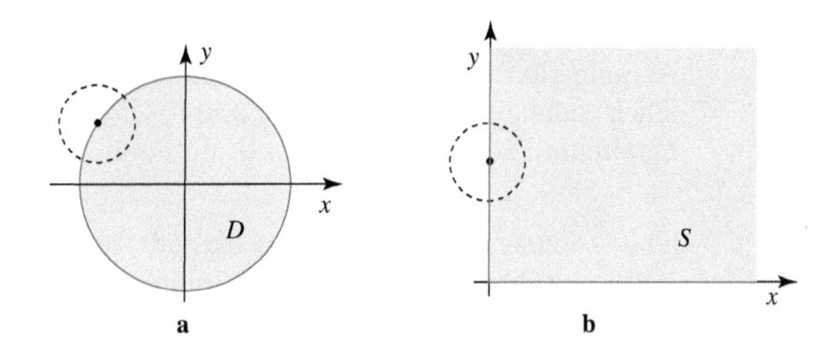

FIGURE 10.17
Closed sets

The set $S = \{(x, y) \mid -1 \leq x \leq 1 \text{ and } 0 < y < 1\}$ drawn in Figure 10.18a is not closed. For instance, $(0, 1)$ is a boundary point of S but does not belong to it. As well, the set $S = \{(x, y) \mid x > 0 \text{ and } 0 \leq y \leq 3\}$ (see Figure 10.18b) is not closed. It is not closed because it does not contain all of its boundary points (for instance, it does not contain $(0, 2)$) and not because it extends to infinity.

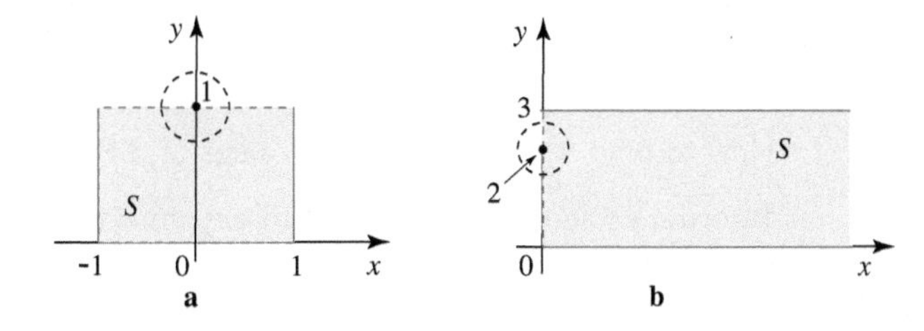

FIGURE 10.18
Sets that are not closed

Definition 23 **Bounded Set**

A set S in $\mathbb{R}^2$ is *bounded* if it is contained within some open disk.

For instance, the set $S = \{(x, y) \mid 0 \leq x \leq 1 \text{ and } 0 \leq y \leq 1\}$ is bounded because it can be placed inside the open disk of radius 2 centred at the origin; see Figure 10.19a. As well, the set in Figure 10.19b is bounded.

On the other hand, the set $S = \{(x, y) \mid x \geq 0 \text{ and } y \geq 0\}$ is not bounded, since no open disk can contain all of S; Figure 10.19c. In general, if a set extends to infinity, then it cannot be bounded.

Note that "bounded" has nothing to do with "boundary."

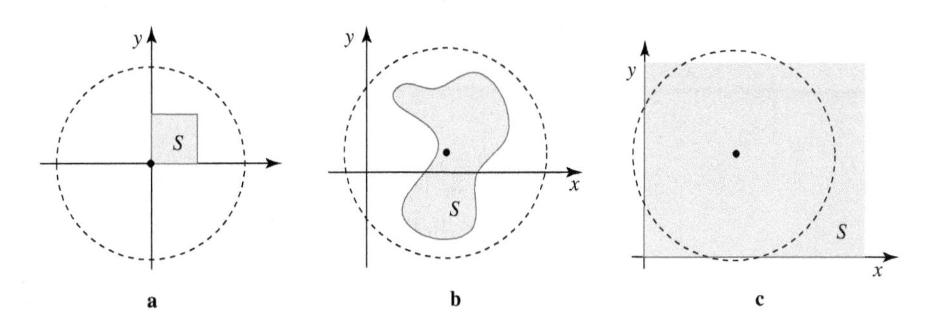

FIGURE 10.19
Bounded and unbounded sets

Definition 24 **Absolute Extreme Values**

If $f(a, b) \geq f(x, y)$ for all (x, y) in a set S, then we say that f has an *absolute* (or *global*) *maximum* on S at (a, b). If $f(a, b) \leq f(x, y)$ for all (x, y) in S, then f has an *absolute* (or *global*) *minimum* on S at (a, b).

Absolute maximum and absolute minimum values are referred to as *absolute extreme values,* or *absolute extrema.*

We are now ready to state our main result.

Theorem 21 Extreme Value Theorem for Functions of Two Variables

Assume that a function f is continuous on a closed and bounded set S in $\mathbb{R}^2$. Then there exist points (a_1, b_1) and (a_2, b_2) in S such that $f(a_1, b_1)$ is an absolute maximum and $f(a_2, b_2)$ is an absolute minimum of f on S.

The statement of the theorem extends to functions of any number of variables (with suitable definitions of continuity, closedness, and boundedness). As is its one-variable counterpart, Theorem 21 is an existence statement: it guarantees that the absolute maximum and minimum exist but does not tell us how to find them or at how many points these values occur.

All assumptions in the theorem are essential: if any one is removed, the theorem may no longer be true (see Exercises 37 and 38).

According to Theorem 18, if f has an extreme value at a point (a, b) inside S (i.e., not on its boundary), then (a, b) must be a critical point. Thus, the absolute extreme values must occur either at critical points in S or at points on the boundary of S. To summarize:

Algorithm 1 How to Find Absolute Extreme Values of a Function

In order to identify the absolute extreme values of a continuous function f on a closed and bounded set S:

(a) Compute the values of f at all critical points in S.

(b) Find the extreme values of f on the boundary of S.

(c) Select the largest and the smallest values obtained in (a) and (b).

Example 10.12 Finding Absolute Extreme Values

Identify the absolute maximum and minimum values of the function $f(x, y) = xy^2 - x - 2y$ on the set $S = \{(x, y) \mid 0 \le x \le 2 \text{ and } 0 \le y \le 5\}$.

Note that f is a polynomial (hence continuous everywhere, and in particular on S) and that the set S is closed and bounded.

First, we find critical points of f. From

$$f_x = y^2 - 1 = 0$$

we get $y^2 = 1$ and $y = \pm 1$. From

$$f_y = 2xy - 2 = 0$$

we get $xy - 1 = 0$. Substituting $y = 1$ into $xy - 1 = 0$ gives $x = 1$. Likewise, when $y = -1$ then $xy - 1 = 0$ implies that $x = -1$. Since the first partials are defined everywhere in $\mathbb{R}^2$, we conclude that $(1, 1)$ and $(-1, -1)$ are the only critical points.

Furthermore, only $(1, 1)$ belongs to S; the value of the function is $f(1, 1) = -2$. (This concludes step (a) from Algorithm 1.)

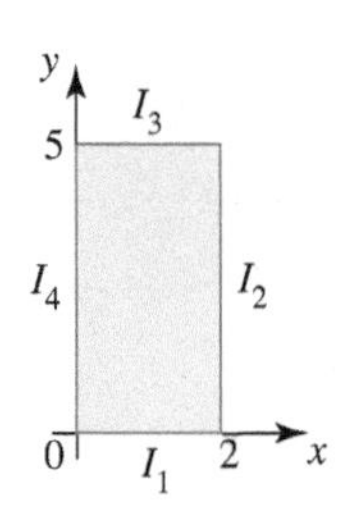

FIGURE 10.20
The set S from Example 10.12

The boundary of S consists of four line segments (Figure 10.20).

The segment l_1 is given by $y = 0$ and $0 \le x \le 2$. On it, f is equal to

$$f(x, 0) = x(0)^2 - x - 2(0) = -x$$

So we need to find the extreme values of the function of one variable $f(x, 0) = -x$ on the interval $[0, 2]$. Since $f(x, 0)$ is decreasing, its extreme values occur at the ends of the interval. The maximum value is $f(0, 0) = 0$ and the minimum value is $f(2, 0) = -2$.

We need to repeat this kind of analysis for the remaining three boundary line segments. It will take a bit of time to do so but it is not difficult; as we have just seen, on the boundary, the problem reduces to finding extreme values of functions of *one* variable.

The line segment I_2 is given by $x = 2$ and $0 \le y \le 5$; this time f is equal to

$$f(2, y) = (2)y^2 - (2) - 2y = 2(y^2 - y - 1)$$

We now employ the technique of finding absolute extreme values of functions of one variable to the function $f(2, y) = 2(y^2 - y - 1)$ on the interval $[0, 5]$.

From $f(2, y)' = 2(2y - 1) = 0$ we get the critical point $y = 1/2$. Comparing

$$f(2, 1/2) = (2)(1/2)^2 - (2) - 2(1/2) = -5/2$$

with $f(2, 0) = -2$ and $f(2, 5) = 38$ we see that the minimum is $f(2, 1/2) = -5/2$ and the maximum is $f(2, 5) = 38$.

On I_3, $y = 5$, $0 \le x \le 2$, and $f(x, 5) = 25x - x - 10 = 24x - 10$. This is an increasing function; thus $f(0, 5) = -10$ is a minimum and $f(2, 5) = 38$ is a maximum.

Finally, on I_4, $x = 0$ and $0 \le y \le 5$, and $f(0, y) = -2y$. The minimum is $f(0, 5) = -10$ and the maximum is $f(0, 0) = 0$.

We have calculated all values suggested by (a) and (b) in Algorithm 1, and we list them here: $f(1, 1) = -2$, $f(0, 0) = 0$, $f(2, 0) = -2$, $f(2, 1/2) = -5/2$, $f(2, 5) = 38$, and $f(0, 5) = -10$. It follows that the absolute maximum of f on S is $f(2, 5) = 38$ and the absolute minimum is $f(0, 5) = -10$.

Figure 10.21 shows the level curves of f within S, suggesting the locations of absolute extreme values. ▲

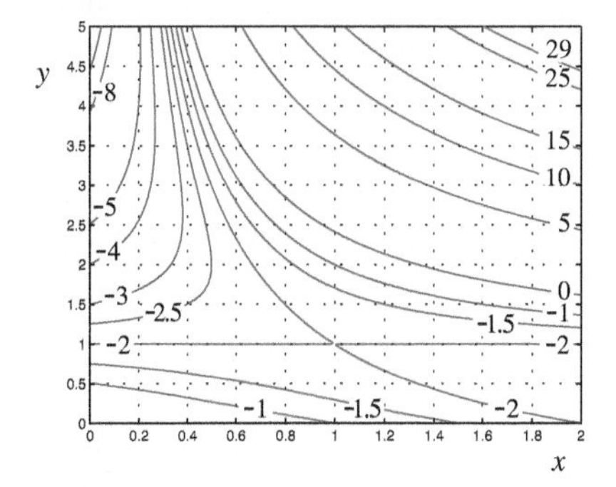

FIGURE 10.21

Level curves of $f(x, y) = xy^2 - x - 2y$

Example 10.13 Finding Extreme Values

Identify the absolute maximum and minimum values of $f(x, y) = \sqrt{x^2 + y^2}$ on the set S in Figure 10.22.

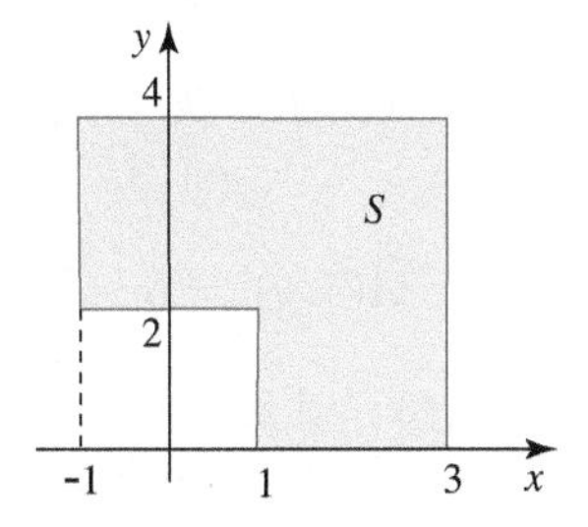

FIGURE 10.22

The set of Example 10.13

Note that f is continuous and S is a closed and bounded set.

One way to do this is to repeat the lengthy procedure that we used in Example 10.12. In this case, however, there is a way to do it that is a lot faster. The idea is to recognize that f has a geometric meaning: it measures the distance from the origin to a point (x, y) in the plane.

Thus, the absolute minimum occurs at the point in S that is closest to the origin, namely $(1, 0)$. The absolute minimum value is $f(1, 0) = 1$. The absolute maximum occurs at the point in S that is farthest from the origin—in this case, it is $(3, 4)$. Thus, $f(3, 4) = \sqrt{3^2 + 4^2} = 5$ is the absolute maximum value of f on the given set.

Example 10.14 Building an Optimal Box with Further Restrictions

Consider building an optimal box as in Example 10.10, but with an additional requirement: the length and the width of the box cannot be smaller than 1 cm or larger than 3 cm. In other words, we are asked to find the minimum of the function

$$A(x, y) = 2xy + \frac{128}{x} + \frac{128}{y}$$

on the set $S = \{(x, y) \mid 1 \leq x \leq 3 \text{ and } 1 \leq y \leq 3\}$.

Recalling the symmetry principle we briefly discussed following Example 10.10, we predict that the optimal box will be the most symmetric given the restrictions. We cannot make a cube this time, but at least we can make a box with a square base.

Note that A is continuous on S and S is closed and bounded.

The only critical point of A, $(4, 4)$, is not in S. So the minimum of A occurs somewhere on the boundary of S. We examine each of the four boundary line segments; see Figure 10.23.

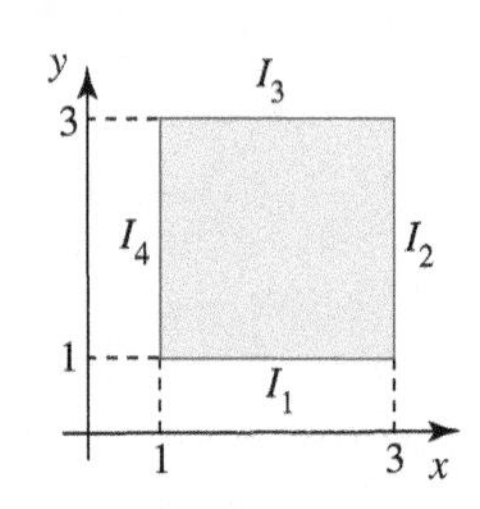

FIGURE 10.23

The set S of Example 10.14

Along I_1, the variables satisfy $y = 1$ and $1 \leq x \leq 3$, and thus

$$A(x, 1) = 2x + \frac{128}{x} + 128$$

From

$$A(x, 1)' = 2 - \frac{128}{x^2} = 0$$

we get $128/x^2 = 2$, $x^2 = 64$, and so $x = \pm 8$. Neither value belongs to $[1, 3]$, so we ignore them; thus, the extreme values occur at the endpoints:

$$A(1, 1) = 2 + 128 + 128 = 258$$

and

$$A(3, 1) = 6 + 128/3 + 128 = 530/3 \approx 176.7$$

Along I_2, the variables satisfy $x = 3$ and $1 \leq y \leq 3$, and A is equal to

$$A(3, y) = 6y + \frac{128}{3} + \frac{128}{y}$$

From

$$A(3, y)' = 6 - \frac{128}{y^2} = 0$$

we get $128/y^2 = 6$, $y^2 = 64/3$ and so $x = \pm 8/\sqrt{3}$. Again, neither critical point is in $[1, 3]$. The candidates for extreme values of A must be the endpoints

$$A(3, 1) = 6 + 128/3 + 128 = 530/3 \approx 176.7$$

and

$$A(3, 3) = 18 + 128/3 + 128/3 = 310/3 \approx 103.3$$

Along I_3, $y = 3$ and $1 \le x \le 3$, and

$$A(x,3) = 6x + \frac{128}{x} + \frac{128}{3}$$

Note that this is the same situation as along I_2. There is no need to proceed, as we will not obtain new candidates for extreme values of A. Along I_4, $x = 1$ and $1 \le y \le 3$, and

$$A(1,y) = 2y + 128 + \frac{128}{y}$$

which is the same function as the one along I_1.

It follows that the absolute minimum is $A(3,3) = 310/3$. The dimensions of the box are $x = 3$, $y = 3$ and $z = 64/xy = 64/9$, so the box does have a square base, as we predicted. ▲

Appendix: Proof of the Second Derivatives Test

Why does the second derivatives test work? Here, we give a sketch of the proof that will give us some insight into the whole situation with extreme values.

Assume that f has continuous partial derivatives (so that $f_{xy} = f_{yx}$) and that (a,b) is a critical point of f (i.e., $f_x(a,b) = 0$ and $f_y(a,b) = 0$). Recall that for (x,y) near (a,b), $f(x,y) \approx T_2(x,y)$, where $T_2(x,y)$ is the degree-2 Taylor polynomial of f (see Theorem 13 in Section 7). Thus,

$$\begin{aligned} f(x,y) &\approx f(a,b) + f_x(a,b)(x-a) + f_y(a,b)(y-b) \\ &\quad + \frac{1}{2}\left(f_{xx}(a,b)(x-a)^2 + 2f_{xy}(a,b)(x-a)(y-b) + f_{yy}(a,b)(y-b)^2\right) \\ &= f(a,b) + \frac{1}{2}\left(f_{xx}(a,b)(x-a)^2 + 2f_{xy}(a,b)(x-a)(y-b) + f_{yy}(a,b)(y-b)^2\right) \end{aligned}$$

since $f_x(a,b) = 0$ and $f_y(a,b) = 0$.

To simplify the calculations we shift the graph of f so that the critical point (a,b) moves to the origin. In that case,

$$f(x,y) \approx f(0,0) + \frac{1}{2}\left(f_{xx}(0,0)x^2 + 2f_{xy}(0,0)xy + f_{yy}(0,0)y^2\right)$$

for (x,y) near $(0,0)$. Let $A = f_{xx}(0,0)$, $B = f_{xy}(0,0)$, and $C = f_{yy}(0,0)$ so that

$$f(x,y) \approx f(0,0) + \frac{1}{2}\left(Ax^2 + 2Bxy + Cy^2\right) \tag{10.2}$$

Formula (10.2) says something important: *near* a critical point, the graph of a function $f(x,y)$ (whose second partials are continuous) looks like the graph of a second-degree polynomial in two variables.

Next, we complete the square to further analyze (10.2):

$$\begin{aligned} Ax^2 + 2Bxy + Cy^2 &= A\left[x^2 + \frac{2B}{A}xy + \frac{C}{A}y^2\right] \\ &= A\left[\left(x + \frac{B}{A}y\right)^2 - \left(\frac{B}{A}y\right)^2 + \frac{C}{A}y^2\right] \\ &= A\left[\left(x + \frac{B}{A}y\right)^2 - \frac{B^2y^2}{A^2} + \frac{Cy^2}{A}\right] \\ &= A\left[\left(x + \frac{B}{A}y\right)^2 + \frac{AC - B^2}{A^2}y^2\right] \end{aligned}$$

Going back to (10.2),

$$f(x,y) - f(0,0) \approx \frac{A}{2}\left[\left(x + \frac{B}{A}y\right)^2 + \frac{AC - B^2}{A^2}y^2\right] \tag{10.3}$$

The first term within the square brackets is always positive. Thus, if $AC - B^2 > 0$ then the whole expression within the square brackets is positive. Assuming that $A < 0$, we conclude that

$$f(x,y) - f(0,0) \approx \frac{A}{2}\left[\left(x + \frac{B}{A}y\right)^2 + \frac{AC - B^2}{A^2}y^2\right] < 0,$$

i.e.,

$$f(x,y) \leq f(0,0)$$

for (x,y) near $(0,0)$. Thus, $f(0,0)$ is a local maximum.

Note that

$$AC - B^2 = f_{xx}(0,0)f_{yy}(0,0) - (f_{xy}(0,0))^2$$

is the function D from the second derivatives test evaluated at $(0,0)$. As well, $A = f_{xx}(0,0)$, so we just proved the statement (a) from Theorem 19. Statement (b) is proven analogously.

If $AC - B^2 < 0$, then the expression within the square brackets in (10.3) is both positive and negative (see Exercise 40 for details) and thus $f(x,y) - f(0,0)$ is both positive and negative. So, near $(0,0)$, $f(x,y) > f(0,0)$ at some points (x,y), and $f(x,y) < f(0,0)$ at some points (x,y): thus, $(0,0)$ is a saddle point.

Rewriting (10.3) as

$$f(x,y) \approx f(0,0) + \frac{A}{2}\left[\left(x + \frac{B}{A}y\right)^2 + \frac{AC - B^2}{A^2}y^2\right]$$

we see that the level curves of f are approximately given by

$$f(0,0) + \frac{A}{2}\left[\left(x + \frac{B}{A}y\right)^2 + \frac{AC - B^2}{A^2}y^2\right] = K$$

where K is a constant. Simplifying, we get

$$\left(x + \frac{B}{A}y\right)^2 + \frac{AC - B^2}{A^2}y^2 = \frac{2}{A}(K - f(0,0)) \tag{10.4}$$

When $AC - B^2 > 0$, then both terms on the left side have the same sign, and (10.4) represents an ellipse. In other words, *near* a local minimum or a local maximum point, the level curves look approximately like ellipses (see Figures 10.8, 10.9, and 10.10). If $AC - B^2 < 0$, the two terms on the left side of (10.4) are of opposite signs, and we obtain a hyperbola. Thus, *near* a saddle point, the level curves look approximately like hyperbolas (see Figures 10.8 and 10.11).

Summary

A function has a **local maximum (minimum)** at a point in its domain if it assumes no larger (smaller) values nearby. To find local extreme values, we identify **critical points,** and then apply the **second derivatives test.** In the cases where the test fails, we use algebraic or geometric reasoning to figure out what is going on. As an application, we constructed the **regression line,** which can be used to describe the data points that tend to cluster around a line. A continuous function defined on a closed and bounded set has **absolute extreme values.** To identify them, we examine all critical points within the set, as well as the points on its boundary.

10 Exercises

1. Draw a contour diagram of a function that has no local extreme values.

2. Draw a contour diagram of a function that has a minimum at $(-1, 0)$ and a saddle point at $(1, 1)$.

3. Reason geometrically (i.e., without the second derivatives test) to show that the function $f(x, y) = x^4 - 4xy^2$ has a saddle point at $(0, 0)$.

4. Assume that a differentiable function f has a local maximum at (a, b). Define a function $g(y) = f(a, y)$ and use it to prove that $f_y(a, b) = 0$.

5. Explain why the set $S = \{(x, y) \mid 0 \leq x \leq 1 \text{ and } y > 0\}$ is not bounded.

6. Is the set $S = \{(x, y) \mid 0 < x^2 + y^2 \leq 1\}$ closed?

7–10 ▪ Looking at the gradient vector field of a differentiable function f, identify locations (if any) where f has a local minimum, a local maximum, or a saddle point.

7.

8.

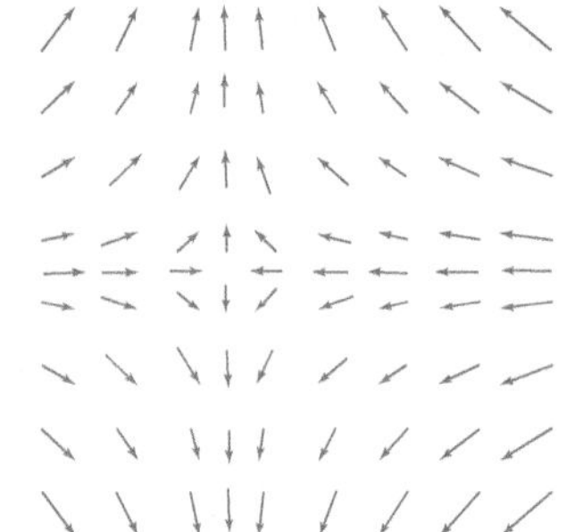

9.

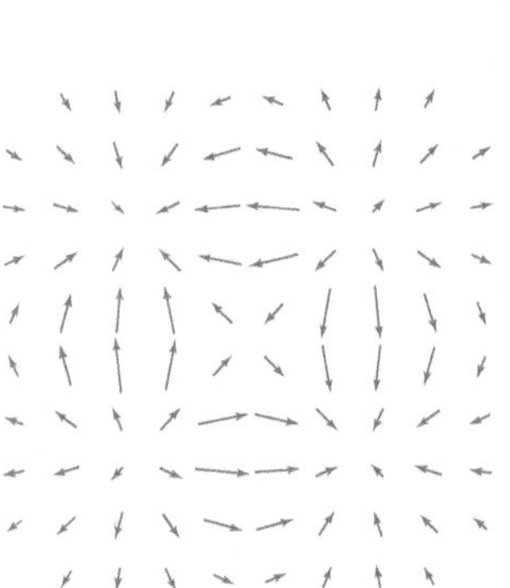

10.

11. Use the second derivatives test to show that $z = x^2 - y^2$ has a saddle point at $(0, 0)$, thus confirming the conclusion we made in Example 10.7.

12–21 ▪ Find the local minimum and maximum values and saddle points (if any) of each function.

12. $f(x, y) = x^2 + y^2 + 2xy^2$

13. $f(x, y) = x^2 - 5xy - y^2$

14. $f(x, y) = x^3 - 2y^2 + 3xy + 4$

15. $f(x, y) = xye^{-x^2-y^2}$

16. $f(x, y) = xye^{-x-y}$

17. $f(x, y) = (x^3 - x)e^{-y^2}$

18. $f(x, y) = e^x \sin y$

19. $f(x, y) = x \cos y$

20. $f(x, y) = x + \dfrac{x+y}{xy}$

21. $f(x, y) = x + y + \dfrac{1}{xy}$

22. Find the dimensions of a closed, rectangular box of given volume $V > 0$ that has minimum surface area.

23. Let $f(x,y)=\sqrt{x^2+y^2}$. Use the definition of the partial derivatives to show that $f_x(0,0)$ and $f_y(0,0)$ do not exist (i.e., are not real numbers).

24. Find all local minimum and maximum values of the function $f(x,y)=9x^2-6xy+y^2+4$.

25. The Shannon index (also known as the Shannon-Wiener index) can be used to measure the diversity of species living in an ecosystem. In the case of three species, it is defined by the formula $H = -a\ln a - b\ln b - c\ln c$, where a is the percentage of species A in the ecosystem, b is the percentage of species B in the ecosystem, and c is the percentage of species C in the ecosystem.

    (a) Using $a+b+c=1$ (why is this true?), reduce H to a function of two variables.

    (b) What is the maximum value of H and when does it occur? Explain what it means for the diversity of species in the given ecosystem.

26. Simpson's diversity index is a common way of assessing the diversity of species living in a region. (Although it is usually used for studying vegetation, the index has been applied to animals as well.) In the case of three species, it is defined by

$$D = 1 - a^2 - b^2 - c^2$$

where a is the percentage of species A in the ecosystem, b is the percentage of species B in the ecosystem, and c is the percentage of species C in the ecosystem. Find the maximum value of D and explain its meaning. (Hint: Use the fact that $a+b+c=1$ to reduce D to a function of two variables.)

27–32 ▪ Find the absolute minimum and maximum values of each function on the given set.

27. $f(x,y)=xy-3x+y$; $S=\{(x,y) \mid -2\le x\le 1 \text{ and } 0\le y\le 3\}$

28. $f(x,y)=x^2+y^2-2$; $S=\{(x,y) \mid 0\le x\le 1 \text{ and } 0\le y\le 2\}$

29. $f(x,y)=xe^y$; $S=\{(x,y) \mid -1\le x\le 1 \text{ and } -3\le y\le 3\}$

30. $f(x,y)=x^2-3x+y$; S is the triangular region with vertices $(0,0)$, $(1,0)$, and $(0,1)$

31. $f(x,y)=\ln(x^2+y^2+1)$; $S=\{(x,y) \mid 0\le x\le 1 \text{ and } 0\le y\le 1\}$

32. $f(x,y)=xy+4$; S is the triangular region with vertices $(0,0)$, $(2,0)$, and $(2,2)$

33. The temperature at a point (x,y) on a metal plate S in the shape of the square $\{(x,y) \mid 0\le x\le 1, 0\le y\le 1\}$ is given by $T(x,y)=2x^2e^y$. Identify the warmest and coldest points in S.

34. The pressure at a point (x,y) on a membrane in the shape of the disk $\{(x,y) \mid x^2+y^2\le 1\}$ is given by $p(x,y)=e^{x^2+y^2+2}$. Find the points on the membrane where the pressure is the strongest and where it is the weakest.

35. Consider the functions $f_1(x,y)=x^4+y^4$, $f_2(x,y)=14-x^4-y^4$ and $f_3(x,y)=2x^4-y^4$.

    (a) Show that $(0,0)$ is the only critical point of each function.

    (b) Show that the second derivatives test cannot be applied to analyze $(0,0)$ for the existence of extreme values.

    (c) Using alternative arguments, determine whether $(0,0)$ is a local minimum, a local maximum, or a saddle point for each of the three functions.

36. Consider the functions $f_1(x,y)=x^2$, $f_2(x,y)=3-x^2$, and $f_3(x,y)=x^3$.

    (a) Find the critical points of each function.

    (b) Show that, in each case, $D(x,y)=0$.

    (c) Sketch the graphs of the three functions, and use them to determine what happens at their critical points.

37. Consider the functions $f_1(x, y) = \sqrt{x^2 + y^2}$ and $f_2(x, y) = \sin x \sin y$, both defined on the set $S = \{(x, y) \mid -2 < x < 2 \text{ and } -2 < y < 2\}$.

    (a) Show that f_1 does not have an absolute maximum on S. Why does this fact not violate the conclusions of Theorem 21?

    (b) Show that f_2 has an absolute maximum and an absolute minimum on S.

    (c) Explain what (a) and (b) mean for Theorem 21.

38. Consider the functions

$$f_1(x, y) = \begin{cases} x^2 + y^2 & \text{if } (x, y) \neq (0, 0) \\ 1/2 & \text{if } (x, y) = (0, 0) \end{cases} \quad \text{and} \quad f_2(x, y) = \begin{cases} x^2 + y^2 & \text{if } (x, y) \neq (0, 0) \\ -1/2 & \text{if } (x, y) = (0, 0) \end{cases}$$

both defined on $S = \{(x, y) \mid x^2 + y^2 \leq 1\}$.

    (a) Show that f_1 does not have an absolute minimum on S. Why does this fact not violate the conclusions of Theorem 21?

    (b) Show that f_2 has an absolute maximum and an absolute minimum on S.

    (c) Explain what (a) and (b) mean for Theorem 21.

39. Find all critical points of the function $f(x, y) = |x| + |y|$.

40. Consider the function $B(x, y) = (x + By/A)^2 + (AC - B^2)y^2/A^2$ appearing within the square brackets in (10.3). Assume that $AC - B^2 < 0$. Using the suggested steps, show that, no matter what open disk around $(0, 0)$ we take, $B(x, y)$ will be positive for some points in the disk and negative for some other points.

    (a) Show that $B(x, y) < 0$ along the points on the line $y = -Ax/B$ (assume that $B \neq 0$).

    (b) Assume that $B = 0$. Find a line that goes through $(0, 0)$ such that $B(x, y) < 0$ at points on that line.

    (c) Find a line that goes through $(0, 0)$ such that $B(x, y) > 0$ at points on that line.

    (d) Explain how your answers to (a) to (c) complete the required proof.

41–44 ▪ Find the regression line for each set of data points.

41. $(-2, 2)$, $(1, 1)$, $(0, 4)$

42. $(0, 0)$, $(0, 1)$, $(1, 1)$

43. $(0, 0)$, $(0, 1)$, $(1, 1)$, $(2, 2)$

44. $(-3, 2)$, $(-1, 1)$, $(1, 0)$, $(1, 1)$

45. In an experiment, we obtained n data points $(x_1, y_1), (x_2, y_2), \ldots, (x_n, y_n)$ that we would like to approximate by a line. So let $y = mx + b$ be the regression line (i.e., the line that best fits the given data in the sense explained in Example 10.11). Show that m and b satisfy the following equations:

$$m(x_1 + x_2 + \cdots + x_n) + bn = y_1 + y_2 + \cdots + y_n$$
$$m(x_1^2 + x_2^2 + \cdots + x_n^2) + b(x_1 + x_2 + \cdots + x_n) = x_1y_1 + x_2y_2 + \cdots + x_ny_n$$

46. Using the result of Exercise 45, find the regression line for the data points $(-3, 2)$, $(-1, 1)$, $(1, 0)$, and $(1, 1)$.

11 Optimization with Constraints

In Section 10, we studied two types of optimization problems: first, we looked for points (x, y) anywhere in the domain where a function has a local minimum or a local maximum. Next, we learned how to identify absolute extreme values on a closed and bounded subset of the domain of a function (which, in many cases, is a more realistic situation).

In this section we attempt to find extreme values of a function whose **independent variables satisfy an additional condition,** given in the form of an equation. This equation (called the *constraint*) defines a subset of the domain, and we look for extreme values on that subset.

Here is an example: suppose that a function $c(x, y)$ models the concentration of a nutrient within a cell and in the surrounding area (Figure 11.1a). A question we asked in Section 10 is, find all location(s) where the concentration of the nutrient is the highest. These locations could be inside a cell, outside of it, or in its membrane. (This is what we call *unconstrained optimization.*) Now, we ask the following question: where, within the membrane of the cell, is the concentration the highest? (This is an example of a *constrained optimization* problem.)

Assume that we model the cell by the disk $\{(x, y) \mid x^2 + y^2 \leq 1\}$ and its membrane by the circle $\{(x, y) \mid x^2 + y^2 = 1\}$ (Figure 11.1b). The question we are asking is, find the maximum value of $c(x, y)$ subject to $x^2 + y^2 = 1$.

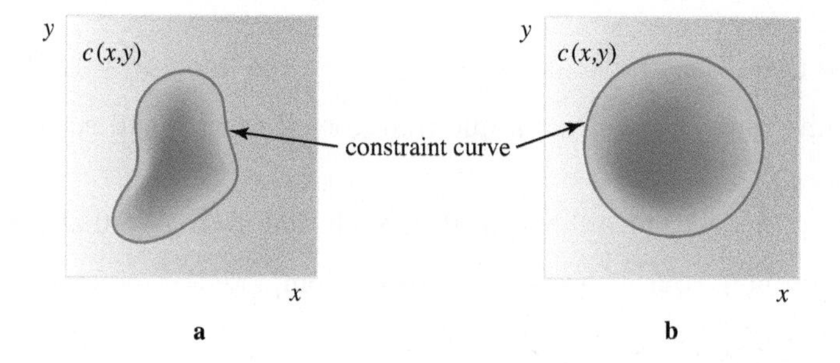

FIGURE 11.1
Constrained optimization problem

Example 11.1 Constrained Optimization Problem

We need to build a closed cylindrical tank of maximum possible volume given that we have $A = 24\pi$ m^2 of material available. Describe this problem as an optimization problem with a constraint.

▶ Denote by r and h the radius and the height, respectively, of the tank. We are asked to find the maximum value of the volume

$$V(r, h) = \pi r^2 h$$

Note that, without any constraints, the problem makes no sense: by increasing the radius, or the height, or both, we can build a tank that will hold any volume, as large as we desire. (In other words, the function $V(r, h) = \pi r^2 h$ does not have a local maximum.)

In our case, the values for r and h are restricted by the requirement on the amount of material that we are allowed to use. The surface area of the tank is (top disk + bottom disk + side)

$$\pi r^2 + \pi r^2 + 2\pi rh = 2\pi r^2 + 2\pi rh$$

and the requirement is that $2\pi r^2 + 2\pi rh = 24\pi$.

Thus, we are asked to find the maximum value of the function $V(r, h) = \pi r^2 h$ subject to the constraint $2\pi r^2 + 2\pi rh = 24\pi$.

We will solve this problem in Example 11.5. ▲

We now state what a constrained optimization problem is, and then reason geometrically to solve it.

Constrained optimization problem: Assume that a function $f(x, y)$ has continuous partial derivatives. Find minimum and maximum values of f subject to the constraint $g(x, y) = k$, where g has continuous partial derivatives and k is some real number.

The constraint $x^2 + y^2 = 1$ from our introductory example can be written as $g(x, y) = 1$, where $g(x, y) = x^2 + y^2$. In Example 11.1 the constraint is $g(r, h) = 2\pi r^2 + 2\pi rh = 24\pi$.

Represent the function f by its contour diagram. The constraint $g(x, y) = k$ is a curve in the xy-plane; see Figure 11.2.

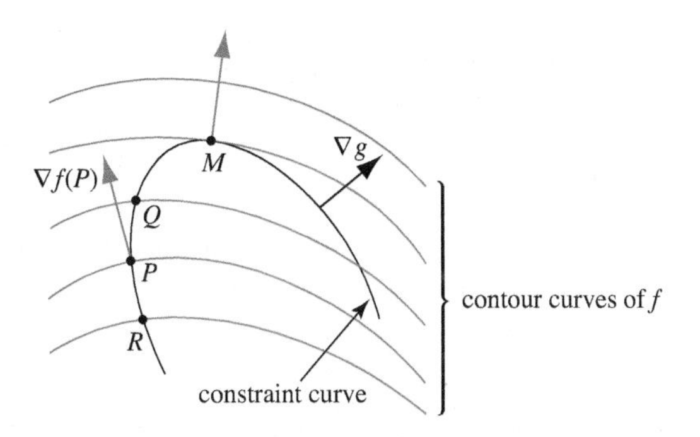

FIGURE 11.2

Investigating optimization with a constraint

As we walk along the constraint curve, we cross the contour curves of f. Suppose that we are at the point P and that the gradient of f is nonzero there, i.e., $\nabla f(P) \neq \mathbf{0}$. Recall that the directional derivative of f is given by

$$D_{\mathbf{u}}f(P) = \|\nabla f(P)\| \cos\theta$$

where θ is the angle between the direction $\mathbf{u}$ and the gradient $\nabla f(P)$ (see formula (9.4) in Section 9). Note that $D_{\mathbf{u}}f(P) > 0$ if $0 \leq \theta < \pi/2$; thus, f increases in any direction that makes an angle of less than $\pi/2$ with respect to the gradient $\nabla f(P)$. (If the angle is greater than $\pi/2$, then $D_{\mathbf{u}}f(P) < 0$ and f is decreasing.)

If we move along $g(x, y) = k$ from P to Q, the function f increases (since this direction makes an angle smaller than $\pi/2$ with respect to the gradient of f), so f does not have a maximum at P. Likewise, if we walk from P to R, f decreases, so P cannot be a minimum either.

We have reached an important conclusion: as long as the gradient vector ∇f at some point is not perpendicular to the constraint curve $g(x, y) = k$, we can make f larger (or smaller) by walking away from that point.

So it seems that if the gradient vector ∇f is (non-zero and) perpendicular to the constraint curve, then f could have an extreme value subject to the given constraint. In Figure 11.2 the gradient ∇f is perpendicular to $g(x, y) = k$ at M.

Recall that $g(x, y) = k$ can be viewed as a contour curve of the function $z = g(x, y)$ of value k. Thus, ∇g, if non-zero, is perpendicular to the constraint curve $g(x, y) = k$. Since at M, both ∇f and ∇g are perpendicular to the constraint curve, they must be parallel. Thus, there is a real number $\lambda \neq 0$ such that

$$\nabla f(M) = \lambda \nabla g(M)$$

Theorem 22 Lagrange Multipliers

Assume that the functions $f(x, y)$ and $g(x, y)$ have continuous partial derivatives. If f has a local extreme value subject to the constraint $g(x, y) = k$ at (a, b), and if $\nabla g(a, b) \neq \mathbf{0}$, then $\nabla f(a, b) = \lambda \nabla g(a, b)$ for a non-zero real number λ.

The number λ is called a *Lagrange multiplier.* This theorem is the basis for the *method of Lagrange multipliers* that we now outline.

Algorithm 2 How to Find Extreme Values of a Function with a Constraint

(1) Identify all points (a, b) such that

(a) $\nabla f(a, b) = \lambda \nabla g(a, b)$ and $g(a, b) = k$

(b) $\nabla g(a, b) = \mathbf{0}$ and $g(a, b) = k$

(c) (a, b) is an endpoint of the curve $g(x, y) = k$

(2) Assume that the constraint curve is a closed and bounded set. The Extreme Value Theorem (Theorem 21 in Section 10) guarantees that f has a minimum and a maximum somewhere on the constraint curve. So the largest (smallest) of the values $f(a, b)$ for all points identified in (a) to (c) is the maximum (minimum) of f subject to $g(x, y) = k$.

We need to include (b) in Algorithm 2 because that case is missed by Theorem 22. Because extreme values can occur at endpoints, we need to include (c). If $g(x, y) = k$ is not closed, or not bounded, we need to use additional arguments to figure out what is going on (see Examples 11.3 and 11.5).

When solving for points (a, b), it is not necessary to find the values for λ.

Example 11.2 Constrained Optimization

Find the maximum and minimum values of $f(x, y) = x + 2y - 2$ subject to the constraint $2x^2 + y^2 = 18$.

▶ Let $g(x, y) = 2x^2 + y^2$; the constraint can be written as $g(x, y) = 18$.

Note that both f and g are polynomials, and thus all of their partial derivatives are continuous.

We compute $\nabla f = \mathbf{i} + 2\mathbf{j}$ and $\nabla g = 4x\mathbf{i} + 2y\mathbf{j}$. The equation $\nabla f = \lambda \nabla g$ implies that

$$\mathbf{i} + 2\mathbf{j} = \lambda(4x\mathbf{i} + 2y\mathbf{j}) = 4x\lambda\mathbf{i} + 2y\lambda\mathbf{j}$$

i.e.,

$$4x\lambda = 1 \quad \text{and} \quad 2y\lambda = 2$$

Assuming that $\lambda \neq 0$, we find $x = 1/4\lambda$ and $y = 1/\lambda$ and substitute into the constraint:

$$\begin{aligned} 2x^2 + y^2 &= 18 \\ 2\frac{1}{16\lambda^2} + \frac{1}{\lambda^2} &= 18 \\ \frac{18}{16\lambda^2} &= 18 \\ 16\lambda^2 &= 1 \\ \lambda^2 &= \frac{1}{16} \end{aligned}$$

and $\lambda = \pm 1/4$. It follows that $x = 1/4\lambda = \pm 1$ and $y = 1/\lambda = \pm 4$. Thus, we obtain two points, $(1,4)$ and $(-1,-4)$.

From $\nabla g = 4x\mathbf{i} + 2y\mathbf{j} = \mathbf{0}$ it follows that $x = 0$ and $y = 0$; i.e., ∇g is zero only at the origin. Since the constraint curve does not go through the origin, it does not contain points where $\nabla g(a,b) = \mathbf{0}$ (checking part (b) of Algorithm 2).

The constraint curve $g(x,y) = 2x^2 + y^2 = 18$ is an ellipse, which has no endpoints, so we do not get any new candidates for extreme values from part (c) of Algorithm 2. Finally, the ellipse is a closed and bounded set, so assumption (2) is satisfied.

We compute $f(1,4) = 7$ and $f(-1,-4) = -11$ and conclude that $f(1,4) = 7$ is the maximum and $f(-1,-4) = -11$ is the minimum of f subject to the given constraint.

In Figure 11.3 we show the level curves of f (parallel lines since f is a linear function) and the constraint curve (ellipse). At the two points $(1,4)$ and $(-1,-4)$ the gradient vectors ∇f and ∇g are parallel (or, equivalently, the contour curve and the constraint curve have a common tangent line).

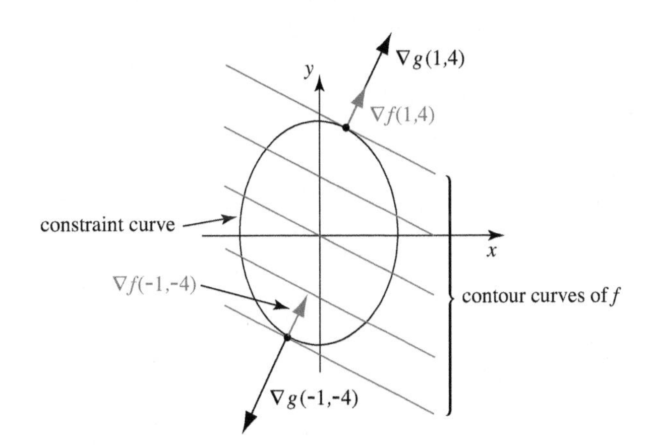

FIGURE 11.3

Visualization of the Lagrange multipliers method

Example 11.3 Constrained Optimization on an Unbounded Curve

Find the maximum and minimum values of $f(x,y) = 3x + y^2$ subject to the constraint $xy = 4$.

The constraint can be written as $g(x,y) = xy = 4$. Both f and g are polynomials, and so their partial derivatives are continuous. From $\nabla f = \lambda \nabla g$ we get

$$3\mathbf{i} + 2y\mathbf{j} = \lambda(y\mathbf{i} + x\mathbf{j}) = \lambda y\mathbf{i} + \lambda x\mathbf{j}$$

and

$$\lambda y = 3 \quad \text{and} \quad \lambda x = 2y$$

Since $xy = 4$ (that's the constraint) neither x nor y can be zero. So, in solving the above system, we can divide by x or y, as needed. Combining $\lambda = 3/y$ and $\lambda = 2y/x$ we get

$$\frac{3}{y} = \frac{2y}{x}$$
$$2y^2 = 3x$$
$$x = \frac{2y^2}{3}$$

Substitute into the constraint equation

$$xy = 4$$
$$\frac{2y^2}{3}y = 4$$
$$y^3 = 6$$
$$y = \sqrt[3]{6}$$

and thus

$$x = \frac{2}{3}y^2 = \frac{2}{3}\sqrt[3]{36}$$

Note that $\nabla g = y\mathbf{i} + x\mathbf{j} = \mathbf{0}$ only when $x = 0$ and $y = 0$. Thus, $\nabla g \neq \mathbf{0}$ everywhere on the constraint curve (that's part (b) of Algorithm 2). The constraint curve $g(x, y) = xy = 4$ is the hyperbola $y = 4/x$, which has no endpoints (part (c)).

Thus, we have only one candidate, $(2\sqrt[3]{36}/3, \sqrt[3]{6})$, for an extreme value of f. At this point,

$$f(2\sqrt[3]{36}/3, \sqrt[3]{6}) = 3\left(\frac{2}{3}\sqrt[3]{36}\right) + \left(\sqrt[3]{6}\right)^2 = 3\sqrt[3]{36} \approx 9.905$$

Is this an extreme value? The constraint curve is a closed set, but it is not bounded (it extends to infinity), so the Extreme Value Theorem does not apply.

The value $f(2\sqrt[3]{36}/3, \sqrt[3]{6})$ is not a maximum, since at the point $(2, 2)$ on the constraint curve

$$f(2, 2) = 3(2) + (2)^2 = 10 > 3\sqrt[3]{36}$$

It is not a minimum, either: the point $(-4, -1)$ lies on the constraint curve, and

$$f(-4, -1) = 3(-4) + (-1)^2 = -11 < 3\sqrt[3]{36}$$

Thus, f has no extreme values subject to the given constraint. ▲

Example 11.4 Constrained Optimization of a Linear Function

Consider the linear function $l(x, y) = Ax + By$, where A and B are fixed numbers such that $A^2 + B^2 \neq 0$. Find the maximum and minimum of l subject to the constraint $x^2 + y^2 = 1$.

▶ Let $g(x, y) = x^2 + y^2$. Note that both f and g have continuous partial derivatives. From $\nabla f = \lambda \nabla g$ we get

$$A\mathbf{i} + B\mathbf{j} = \lambda(2x\mathbf{i} + 2y\mathbf{j}) = 2x\lambda\mathbf{i} + 2y\lambda\mathbf{j}$$

i.e.,

$$2x\lambda = A \quad \text{and} \quad 2y\lambda = B$$

Next, we compute $x = A/2\lambda$ and $y = B/2\lambda$ and substitute into the constraint:

$$x^2 + y^2 = 1$$
$$\frac{A^2}{4\lambda^2} + \frac{B^2}{4\lambda^2} = 1$$
$$\frac{A^2 + B^2}{4\lambda^2} = 1$$
$$\lambda^2 = \frac{A^2 + B^2}{4}$$
$$\lambda = \pm\frac{\sqrt{A^2 + B^2}}{2}$$

Thus,

$$x = \frac{A}{2\lambda} = \pm\frac{A}{2\frac{\sqrt{A^2+B^2}}{2}} = \pm\frac{A}{\sqrt{A^2 + B^2}}$$

and

$$y = \frac{B}{2\lambda} = \pm\frac{B}{\sqrt{A^2 + B^2}}$$

So, we have two candidates for extreme values:

$$(A/\sqrt{A^2 + B^2}, B/\sqrt{A^2 + B^2}) \quad \text{and} \quad (-A/\sqrt{A^2 + B^2}, -B/\sqrt{A^2 + B^2})$$

If $\nabla g = y\mathbf{i} + x\mathbf{j} = \mathbf{0}$, then $x = 0$ and $y = 0$; thus, $\nabla g \neq \mathbf{0}$ everywhere on the constraint curve. The constraint curve $g(x, y) = x^2 + y^2 = 1$ is a circle, which has no endpoints (so we do not get any candidates for extreme values from (b) and (c) in Algorithm 2). The circle is a closed and bounded set, so (2) in Algorithm 2 is satisfied. It follows that

$$\begin{aligned} l(A/\sqrt{A^2 + B^2}, B/\sqrt{A^2 + B^2}) &= A\frac{A}{\sqrt{A^2 + B^2}} + B\frac{B}{\sqrt{A^2 + B^2}} \\ &= \frac{A^2 + B^2}{\sqrt{A^2 + B^2}} \\ &= \sqrt{A^2 + B^2} \end{aligned}$$

is the maximum, and

$$\begin{aligned} l(-A/\sqrt{A^2 + B^2}, -B/\sqrt{A^2 + B^2}) &= A\left(-\frac{A}{\sqrt{A^2 + B^2}}\right) + B\left(-\frac{B}{\sqrt{A^2 + B^2}}\right) \\ &= -\sqrt{A^2 + B^2} \end{aligned}$$

is the minimum of f with the given constraint.

What is the meaning of this calculation? Take a differentiable function f, and let $A = f_x(a, b)$ and $B = f_y(a, b)$ at some point (a, b) in the domain of f. In other words, the vector $A\mathbf{i} + B\mathbf{j}$ is the gradient vector $\nabla f(a, b)$ (the condition $A^2 + B^2 \neq 0$ guarantees it is not zero). Think of x and y as coordinates of a unit vector $\mathbf{u} = x\mathbf{i} + y\mathbf{j}$ ($\mathbf{u}$ is indeed a unit vector, since the constraint requires that $x^2 + y^2 = 1$).

The function $l(x, y)$ can be written as

$$l(x, y) = Ax + By = (A\mathbf{i} + B\mathbf{j}) \cdot (x\mathbf{i} + y\mathbf{j}) = \nabla f(a, b) \cdot \mathbf{u}$$

So $l(x, y)$ is the directional derivative $D_{\mathbf{u}}f(a, b)$ of f at (a, b) thought of as a function of the direction $\mathbf{u} = x\mathbf{i} + y\mathbf{j}$. We have just proved that the maximum of $l(x, y) = D_{\mathbf{u}}f(a, b)$ occurs when

$$\mathbf{u} = x\mathbf{i} + y\mathbf{j} = \frac{A}{\sqrt{A^2 + B^2}}\mathbf{i} + \frac{B}{\sqrt{A^2 + B^2}}\mathbf{j} = \frac{A\mathbf{i} + B\mathbf{j}}{\sqrt{A^2 + B^2}} = \frac{\nabla f(a, b)}{\|\nabla f(a, b)\|}$$

i.e., when $\mathbf{u}$ is the unit vector in the direction of the gradient $\nabla f(a, b)$. The maximum value

$$l(A/\sqrt{A^2 + B^2}, B/\sqrt{A^2 + B^2}) = \sqrt{A^2 + B^2}$$

is the magnitude of the gradient vector, $\|\nabla f(a, b)\|$. The interpretation for the minimum value of l is analogous.

Thus, we re-proved Theorem 15 from Section 9. ▲

Example 11.5 Solution of the Constrained Optimization Problem from Example 11.1

Find the maximum of $V(r, h) = \pi r^2 h$ subject to the constraint $g(r, h) = 2\pi r^2 + 2\pi rh = 24\pi$.

▶ The requirement $\nabla V = \lambda \nabla g$ implies

$$2\pi rh\mathbf{i} + \pi r^2\mathbf{j} = \lambda((4\pi r + 2\pi h)\mathbf{i} + 2\pi r\mathbf{j})$$

and

$$2\pi rh = \lambda(4\pi r + 2\pi h) \quad \text{and} \quad \pi r^2 = 2\pi r\lambda$$

Now we simplify

$$\begin{aligned} rh &= \lambda(2r + h) \\ r &= 2\lambda \end{aligned}$$

and substitute $\lambda = r/2$ into the first equation:

$$\begin{aligned} rh &= \frac{r}{2}(2r + h) \\ 2rh &= 2r^2 + rh \\ rh &= 2r^2 \\ h &= 2r \end{aligned}$$

Using the constraint,

$$\begin{aligned} 2\pi r^2 + 2\pi rh &= 24\pi \\ 2\pi r^2 + 2\pi r(2r) &= 24\pi \\ 6\pi r^2 &= 24\pi \\ r^2 &= 4 \end{aligned}$$

Thus, $r = \pm 2$, but only $r = 2$ makes sense in the context of this problem. It follows that $h = 2r = 4$, and so $(r = 2, h = 4)$ is a candidate for an extreme value of V.

The condition $\nabla g(r,h) = (4\pi r + 2\pi h)\mathbf{i} + 2\pi r\mathbf{j} = \mathbf{0}$ implies that $r = 0$ and $h = 0$; we will not include these values in the list of candidates for extreme values since we need $r, h > 0$.

What is the constraint curve? From

$$g(r,h) = 2\pi r^2 + 2\pi rh = 24\pi$$

we get $r^2 + rh = 12$. Completing the square

$$\left(r + \frac{1}{2}h\right)^2 - \frac{1}{4}h^2 = 12$$

we recognize a hyperbola—it is a closed set, but not bounded (since it extends to infinity); it has no endpoints.

It follows that our only candidate for an extreme value is $(r = 2, h = 4)$; in that case $V(2,4) = \pi(2)^2 4 = 16\pi$. Is this the maximum we have been looking for? Since the constraint curve is not bounded, we cannot use the Extreme Value Theorem, so we need to try something else.

From $r^2 + rh = 12$ it follows that $rh = 12 - r^2$ and $h = 12/r - r$. Thus, as a function of one variable,

$$V(r) = \pi r^2 h = \pi r^2\left(\frac{12}{r} - r\right) = \pi(12r - r^3).$$

From $V'(r) = \pi(12 - 3r^2) = 0$ we get critical points $r = \pm 2$; only $r = 2$ makes sense. Since $V''(r) = -6\pi r$, we get $V''(2) = -12\pi < 0$, and so $V(2)$ is a maximum. We're done. ▲

We used this example as an illustration of how constrained optimization works. As the last part of the solution shows, we did not have to use two variables at all. From the start, we could have eliminated h and analyzed the function $V(r) = \pi(12r - r^3)$ of one variable.

The method of Lagrange multipliers can be extended to functions of three or more variables, as well as to situations that require more than one constraint.

Summary **Optimization with a constraint** involves identifying extreme values of a function whose variables satisfy a condition (called a **constraint**) given in the form of an equation. To identify extreme values, we look for all points on the constraint curve where the **gradient of the constraint** is parallel to the **gradient of the function.** This process is formalized as the **method of Lagrange multipliers.**

11 Exercises

1. Explain geometrically (i.e., by sketching level curves) why $f(x,y) = (x^2+y^2)^{-1}$ cannot have a minimum or maximum subject to the constraint $2x - y = 0$.

2. Minimize the function $f(x,y) = \sqrt{(x-2)^2 + (y-1)^2}$ subject to the constraint $x + y = 0$. Give a geometric interpretation of your answer.

3–6 ▪ Looking at the gradient vector field of each differentiable function f, identify approximate locations (if any) where f has a minimum or a maximum subject to the given constraint curve.

3.

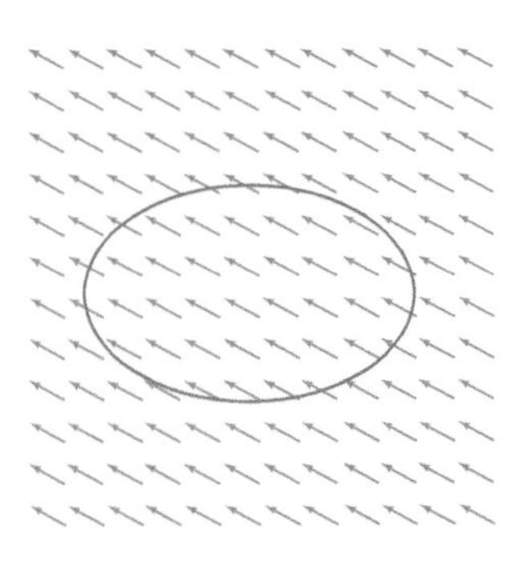

4.

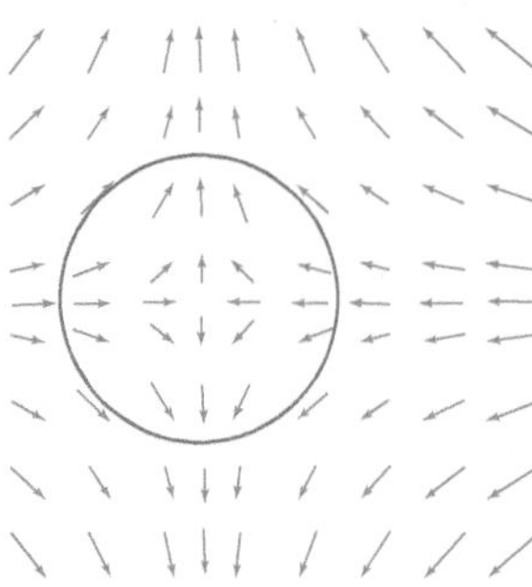

5.

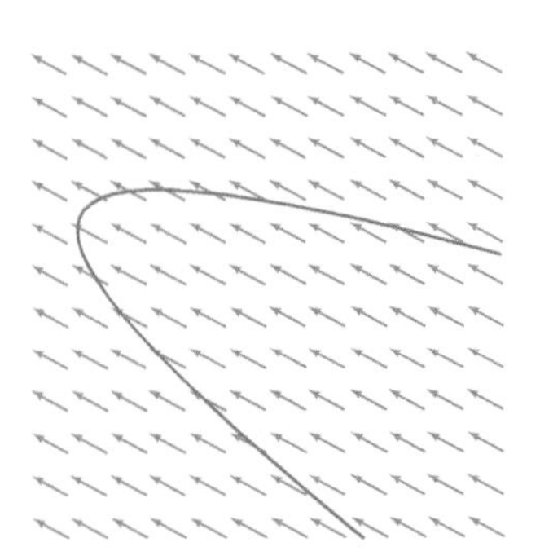

6.

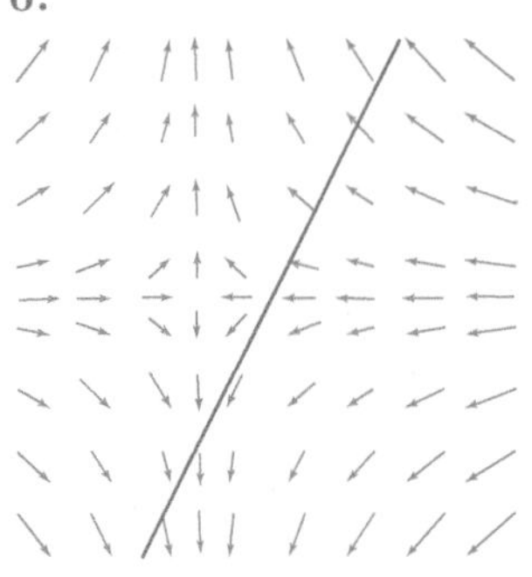

7. Explain why it does not make much sense to use the Lagrange multipliers method to optimize a function of two variables subject to two linear constraints $g_1(x,y) = k_1$ and $g_2(x,y) = k_2$.

8. Sketch the level curves of the function $f(x,y) = x - y + 3$. In the same coordinate system, sketch the graph of the constraint $x^2 + y^2 = 1$.

   (a) In your picture, identify the points where f has a minimum and a maximum subject to the given constraint.

   (b) Using geometric reasoning, find the minimum and maximum of f subject to the given constraint. (Hint: Think about tangent lines.)

   (c) Use the method of Lagrange multipliers to confirm your answers to (a) and (b).

9. Sketch the level curves of the function $f(x,y) = x^2 + y^2$. In the same coordinate system, sketch the graph of the constraint $x^2 + (y-1)^2 = 1$.

(a) In your picture, try to identify the points where f has a minimum and a maximum subject to the given constraint. Can you find them?

(b) Use the method of Lagrange multipliers to find the extreme values of f subject to the given constraint.

10–17 ▪ Use the Lagrange multipliers technique to find the maximum and minimum values (if any) of each function subject to the given constraint.

10. $f(x, y) = 2xy$; $x^2 + y^2 = 9$

11. $f(x, y) = x^3 + y^3$; $x^2 + y^2 = 9$

12. $f(x, y) = \sqrt{x^2 + y^2}$; $xy = 2$ (Hint: Find the minimum, and argue geometrically that the maximum does not exist.)

13. $f(x, y) = x^2 - y^2$; $2x - y = 0$ (Hint: Find the maximum, and argue that the minimum does not exist.)

14. $f(x, y) = x^2y^2$; $3x - y = 1$

15. $f(x, y) = x^2y^2$; $x^2 - y^2 = 1$

16. $f(x, y) = x^2 - 3y^2$; $x^2 + y^2 = 1$

17. $f(x, y) = xy + y$; $x^2 + 2y^2 = 2$

18. Find all points on the curve $xy^2 = 4$ that are closest to the origin.

19. The temperature at a point (x, y) on a metal plate in the shape of the disk $x^2 + y^2 \leq 6$ is $T(x, y) = 2x^2 + y^2 + 120$. Find the coldest point on the rim of the plate.

20. Suppose that you need to build a rectangular box with a lid using 10 m^2 of material. Find the dimensions of the box that holds the largest possible volume.

Index